Naturalists' Handbooks 32

# Ponds and small lakes

BRIAN MOSS

Pelagic Publishing
www.pelagicpublishing.com

Published by Pelagic Publishing
www.pelagicpublishing.com
PO Box 725, Exeter, EX1 9QU, UK

**Ponds and small lakes**
Naturalists' Handbooks 32

*Series Editor*
William D.J. Kirk

ISBN 978-1-78427-135-0 (Pbk)
ISBN 978-1-78427-136-7 (ePub)
ISBN 978-1-78427-137-4 (Mobi)
ISBN 978-1-78427-138-1 (PDF)

Text © Pelagic Publishing 2017

British Library Cataloguing in Publication Data
A catalogue record for this book is available from the British Library

Cover
Top: Bystock Pools nature reserve, Devon. Bottom, left to right: SEM photographs by D.N. Furness, a ciliate (*Coleps* species), a diatom, an amoeba; and by Per Harald Olsen/ NTNU (CC BY 2.0), *Daphnia magna*.

# Contents

# Editor's preface

Nearly everyone has some kind of pond or small lake nearby. It could be a village pond or a pond in a garden, park or nature reserve. They are all around us, but there is still much that we do not understand about them.

Ponds and small lakes support an extremely rich biodiversity of fascinating organisms. Many people have tried pond-dipping and encountered a few unfamiliar creatures, such as dragonfly nymphs and caddisfly larvae. However, there is a far richer world of microscopic organisms, such as diatoms, desmids and rotifers, which is revealed in this book. Anyone with access to a microscope can open up this hidden dimension. Identification keys are provided so that readers can identify, explore and study this microscopic world. There are also many suggestions of ways in which readers can then make original contributions to our knowledge and understanding of pond ecology. It is not even necessary to have access to a pond to be able to study them because artificial ponds that will quickly develop communities of microorganisms can be created easily from jam jars or plastic buckets filled with rainwater.

The book not only explores the fascinating world of the creatures within ponds and their interactions, but also explains the many ways in which ponds are important in human affairs. Ponds are being lost around the world, but they are a key part of a system that maintains our climate. In the face of climate change, it has never been more important to understand the ecology of ponds.

Sadly, Brian Moss, the author, was diagnosed with terminal cancer towards the end of writing the book. Despite this, he continued to work hard on it, knowing his time was limited. Only the acknowledgements were not completed. He passed away in May 2016, before the book was published. Brian's interest, enthusiasm and commitment to ponds live on in this book. I am confident that this book will enable more people to appreciate and investigate these fascinating and important habitats.

William D.J. Kirk
July 2016

# Preface

*Why study a pond?*

*'It's fun' says A, the enthusiastic naturalist.*

*'It's in the syllabus', snaps B, the disillusioned teacher.*

*'Well', says C, the calculating careerist, 'I believe it might throw light on a problem which interests X and Y. They are two coming men, and if I can catch their eye I shall have friends at court in a few years' time when I am looking for a top post'.*

*'A pond presents a limited environment without a continual interchange of population with neighbouring biotopes and is, therefore, suitable for the study of principles,' enunciates D, the serious ecologist.*

*'World starvation is a real threat. In many tropical lands an important amount of protein is raised in ponds. We should study production in ponds wherever we can' explains E, the scientist with a social conscience.*

*We can ignore B and C; A,D and E merit our attention.*

The first paragraph of *Ponds and Lakes* by T.T. Macan, published by Allen and Unwin in 1973.

Brian Moss
April 2016

# About the author

Getting wet and muddy was a childhood trait that Brian Moss never quite grew out of. His research and teaching embraced freshwaters on five continents over fifty years, a range of approaches from field survey to laboratory and whole-lake experiments and a gamut of sites from lakes in Malawi, Tanzania and Michigan, to thermal streams in Iceland, the Norfolk Broads, the North-West Midland Meres and temperature-controlled ponds at the University of Liverpool's Botanic Gardens. When he retired from Liverpool as Professor of Botany in 2008 he was spending at least as much time with invertebrates and fish as with plants and algae.

His work has been widely published, with a textbook, *The Ecology of Freshwaters*, soon to appear in its fifth edition, books in the New Naturalist series on *The Broads* and *Loughs and Lochs*, and a manual on lake restoration. He has been President of the International Society for Limnology and Vice-President of the British Ecological Society.

He was awarded the Institute of Ecology and Environmental Management's annual medal for his life's work and leadership in shallow-lake research in 2010, and the Ecology Institute's 'Excellence in Ecology' prize in 2009. This entailed the writing of a book, *Liberation Ecology*, which interprets ecology for the general public through the media of the fine arts. The book won the Marsh Prize, in 2013, for the best ecology book published in the previous year. Brian loved teaching, playing the double bass (not very well), writing satirical doggerel, often directed at officialdom, and was exercised daily by a large dog.

# About Naturalists' Handbooks

*Naturalists' Handbooks* encourage and enable those interested in natural history to undertake field study, make accurate identifications and make original contributions to research. A typical reader may be studying natural history at sixth-form or undergraduate level, carrying out species/habitat surveys as an ecological consultant, undertaking academic research or just developing a deeper understanding of natural history.

### History of the Naturalists' Handbooks series

The *Naturalists' Handbooks* series was first published by Cambridge University Press, then Richmond Publishing and then the Company of Biologists. In 2010 Pelagic Publishing began to publish new titles in the series together with updated editions of popular titles such as *Bumblebees* and *Ladybirds*. If you are interested in writing a book in this series, or have a suggestion for a good title please contact the series editor.

### About Pelagic Publishing

We publish scientific books to the highest editorial standards in all life science disciplines, with a particular focus on ecology, conservation and environment.

Pelagic Publishing produce books that set new benchmarks, share advances in research methods and encourage and inform wildlife investigation for all.

If you are interested in publishing with Pelagic please contact editor@pelagicpublishing.com with a synopsis of your book, a brief history of your previous written work and a statement describing the impact you would like your book to have on readers.

# 1 Ponds

**Fig. 1.1** The Middle Millpond at Pembroke flanks the mediaeval town.

The great castle of Pembroke is built at the western end of a limestone ridge in south-west Wales. To either side of the ridge, along which the town grew, were two branches of the estuary of the Pembroke River. They ran broadly east-west and discharged ultimately to Milford Haven. In 1946 the southern prong was finally filled in and now forms a piece of land used as roads, car parks and recreational land, with a much engineered stream running through it. The northern prong was dammed, just north of the castle, in the early thirteenth century, to hold a freshwater millpond, the water from which powered a mill set on the dam. The mill burned down and was demolished in 1955 but the dam and its pond (Fig. 1.1) remain and over a couple of decades until 2015 a flock of mute swans on it grew to around 70 birds (Fig. 1.2). The pond has gardens down to the edge on its north side and a masonry embankment, with a walking path paralleling the town wall along the south, so that most of its shoreline is steep or vertical and cannot easily support plants like reeds that would give particularly good habitat for nesting birds and invertebrates. Only at the eastern end, where water enters from another pond, the Upper Millpond, through a narrow tunnel under a railway embankment

**Fig. 1.2** Flocks of mute swans in the Pembroke Millpond grew to such large size as to reduce plant diversity and stimulate large growths of an unattractive alga. The solution was not, however, that apparently favoured by a pondside publican.

was there the possibility of some reed growth though the strength of the westerly winds prevented much developing.

In recent years there has been a move to diversify the habitat through provision of wire mesh mats to encourage growth of reeds along the southern shore and at the western end, but the swans interfered with this. They are voracious feeders on young shoots and their number was boosted by the bread daily fed to them by citizens and tourists. In summer, their grazing diminished growth also of underwater plants so that only a fast growing alga, *Entero-morpha*, looking like stringy lettuce, could keep pace. It was not a very attractive system. In 2015, for unknown reasons, most of the flock decamped elsewhere. The result has been a flourishing growth of the reeds and of a much more diverse plant community in the pond. The algal, invertebrate and fish communities have not been monitored, nor has the water chemistry, so we do not know how they might have changed or why the swans have moved.

There are several lessons from this. First, a heavily modified pond can be made much more attractive, despite concrete banks; secondly, although it is familiar that grazing birds must depend on plant food, the birds, by their grazing, can change the nature of the whole system, just as farm stock can maintain a grassland where otherwise a forest would develop; and thirdly, much information of interest was lost because no-one had time to record properly the plant and invertebrate communities in the pond (although the bird community has been served by several amateur birdwatchers). All I know comes from casual walks around the pond during frequent but short visits. There was potentially an important role for local amateur naturalists, had the opportunity been recognised.

I hope that this book will give insights into the natural history of ponds that might help such opportunities to be taken, and beyond that to bring the enormous fascination and importance of freshwaters, the science of limnology, to a wider audience. Science advances partly on the collection and sifting of information, but mostly on fortuitous or deliberate experiments that can give understanding of mechanisms, so as well as providing advice on sampling and identification of the animals and plants, I also suggest experiments that can be carried out largely with domestic materials and equipment, in a garden and on its pond. If you do not have a pond, or access to one locally, that is no problem. Plastic bowls or buckets offer many possibilities. Any activity carries risks and we now live in a society that has become increasingly unadventurous because of the threat of legal action. Schoolchildren now are rarely allowed close to water without lifejackets, goggles, gloves and close supervision. The risks, however, of drowning, eye and other infections are exceptionally small, especially if common sense is applied. The same is true of the laboratory, or in this context the facilities of a domestic kitchen. My assumptions and those of the publisher are that normal common sense and practical precautions, like covering up skin wounds and abrasions, will be used in carrying out any of the suggested investigations.

Professional limnologists are often asked how a pond differs from a lake. There are many possible definitions (an area of less than 2 ha, with water in it for at least four months a year, is one), all of them arbitrary and to which undermining exceptions can always be found. In the end there are no absolute differences. Ponds are small lakes; lakes are large ponds; they are all bulges of various sizes in drainage systems. They exist in an unbroken range of sizes and they are all linked in systems of runnels, streams and rivers, either on the surface of the land, or under it through the groundwater, in a system that connects the rain and snow ultimately with the ocean. This movement is then made into the water cycle by evaporation, wherever water is exposed, and condensation into rain or snow.

There is thus a continuum of standing bodies of water,

**hectare (ha)**
an area of 10,000 m² or 0.01 km²

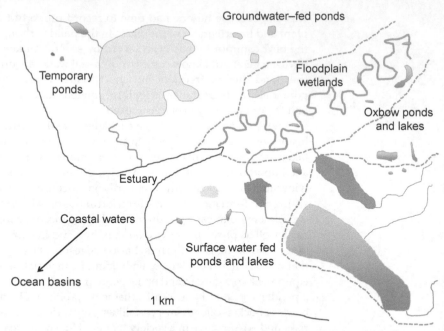

**Fig. 1.3** Landscapes, except where modified by people, have rivers, wetlands and standing waters in a continuous range of sizes and depths; streams and groundwater flows link the whole system through to estuaries, coastal seas and ocean basins

from puddles to ocean basins (Fig. 1.3) and along this series there are steadily changing characteristics that operate in many directions. The mistake is to think in terms of separate categories rather than continuous change. Boundaried thinking has meant that work on large bodies of water, where expensive boats and equipment are needed, has tended to gain more prestige than that on ponds and small lakes, at least until recently. But since the 1970s there has been something of a change in balance, as the importance of shallow wetlands has been realised, and ponds have entered the mainstream in scientific research. Thinking in continua gives a much better idea of how the world works. Moreover, there is no absolute distinction between standing waters (puddles, ponds, lakes and oceans) and flowing waters (runnels, streams, rivers and estuaries), though scientists have tended to think of themselves as lake ecologists or stream ecologists depending on their special interests. There is just a continuum concerning how long, on average, a water molecule might remain around the same spot. It is very short in fast-moving waters, perhaps only a second or two, very long in ocean basins, probably some tens of thousands of years. In puddles it may be moved on by outflow or evaporation in days, in ponds in weeks, in small lakes, months and in large lakes a few years. Slow-flowing rivers on the plains and riverine lakes, where, for example, the river is held back by a gorge through a mountain range have much the same characteristics.

This continuum of retention time, the reciprocal of which is called the turnover rate (the number of times that the water mass is replaced per year), has many consequences. A runnel, where water collects on the land in a temporary channel during a rainstorm, before joining a stream with a permanently recognisable bed, will have water with the chemistry of rain. Streams in areas of hard, poorly weatherable rocks, like granites, and thin soils, will also have the chemistry largely of rain, but in regions where the rain also percolates through porous rocks, deeper soils or peats, the water chemistry will be greatly changed and will reflect the nature of the local geology. The chemistry will change also as a result of the activity of organisms living in the water. Substances are taken up for growth, others are excreted. More changes occur when bodies die and are decomposed. And the longer the retention time, the more these changes will have effect. The volume of the water body will also be important. Small water bodies tend to have a very short retention time. Their water chemistry will be less influenced by their organisms. But retention times beyond a few days will be reflected in much greater changes, many of them caused by the activity of the organisms themselves.

**tarn**
mountain lake or pond in the depression left by a glacier

For example, a small rocky tarn, high on the granite of the mountains, will have water that is barely different from rain, with relatively high sodium and chloride concentrations and low calcium and carbonate concentrations (Fig. 1.4). Indeed it is much diluted seawater because rain picks up droplets of seawater suspended in the atmosphere by storms, as well as wind-blown dust. The closer the tarn to the sea, the greater will be the proportions of sodium and chloride. In contrast, a pond in the lowlands, amid glacial drift soils derived from a myriad of rocks, will have relatively more calcium and carbonate. Moreover the absolute concentrations of calcium and carbonate will be high compared with the mountain tarn. The sodium and chloride concentrations will also be higher, but not much higher. The source of these ions is still largely the rain, though the longer retention times, and higher evaporation in the warmer lowlands, allow some concentration in the local groundwater.

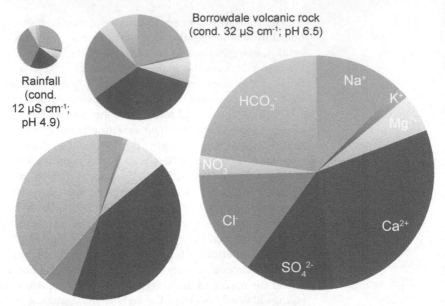

Rainfall
(cond.
12 μS cm⁻¹;
pH 4.9)

Borrowdale volcanic rock
(cond. 32 μS cm⁻¹; pH 6.5)

Sedimentary Carboniferous, Permian &
Triassic rocks (cond. 300 μS cm⁻¹; pH 7.7)

Chalk and soft sedimentary rocks
(cond. 563 μS cm⁻¹; pH 7.6)

**Fig. 1.4** Proportions of the most abundant ions in rainfall, in water from upland tarns on igneous Borrowdale volcanic rocks in the Cumbrian Lake District, in water from tarns on sedimentary Carboniferous, Permian and Triassic rocks in north-west England and in water from ponds, largely on chalky rocks, in south-east England. The colour coding is the same for all and the size of each chart approximately reflects the total amount of ions present. Rainfall is dominated by sodium and chloride, largely derived from sea spray. The water in tarns on igneous rocks strongly reflects the composition of rain with a little more calcium and bicarbonate added. On sedimentary rocks the proportions of sodium and chloride decline whilst calcium and bicarbonate increase. Based on Sutcliffe (1998) and Bennion *et al.* (1997).

**igneous**
formed from cooled lava
or magma

***et al.***
is short for *et alia*
(and others)

**morphometry**
form, shape and size

The next characteristic that is important along the continuum from puddles to ocean basins is morphometry. This describes the shape and depth of the basin. The smaller the basin, the less permanent is the morphometry. The morphometry of ocean basins is determined by the drifting of continents and long-term sedimentation at the edges from erosion of the land. Such basins change very slowly and their main features are very ancient. Less permanent are the deep lake basins formed by large geological movements that contain lakes like Baikal in Russia, and Tanganyika, Malawi, Edward and Turkana in Africa, which are perhaps a few million years old but have nonetheless undergone huge changes in water level, volume and shoreline during their history. Such old basins are very few; most lakes, even the

Area of water body (km²)

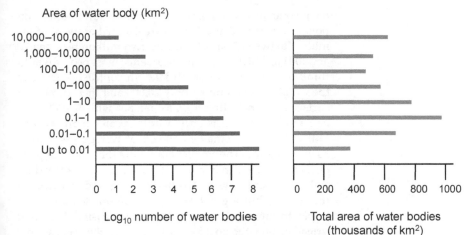

Log₁₀ number of water bodies

Total area of water bodies
(thousands of km²)

**Fig. 1.5** Inspection of detailed maps, or in this case images from satellites, gives a picture of the numbers of water bodies of different sizes and their total areas. Small lakes and ponds from 0.2 ha to 1 ha in area are the most abundant, with about three hundred million of them. However there are probably even more very small ponds, less than 0.2 ha in area, which are not often mapped and not detectable by the satellites used. There are progressively fewer bodies of water as size increases. The largest lakes here (fewer than 100 of them up to 100,000 km²) include the St Lawrence Great Lakes, the East African Great Lakes and Lake Baikal. The total surface water area in each category is much more uniform. Water bodies cover about 5 x 10⁶ km², or about 3.7% of Earth's non-glaciated land area. From Verpoorter *et al.* (2014).

**moraine**

deposit of rock debris, gravel and sand dropped in one place when a glacier melts back at the same rate as the ice is flowing forwards

**mere**

local name for a lake, particularly a small, shallow lake

very large St Lawrence Great Lakes on the USA/Canadian border, the largest lakes of Europe, and every smaller lake and pond including all those of the UK, are very young. Most were formed, one way or another, by the action of ice, and their basins were exposed and filled at most ten to fifteen thousand years ago when the glaciers retreated. Sometimes their floors were bulldozed by the moving ice, sometimes their basins are former river valleys dammed by moraines. Sometimes they formed in great holes left when icebergs calved from the retreating ice front and were buried in glacial drift. The lochs of Scotland, the loughs of Ireland, the llyns of Wales, the lakes of Cumbria and the north-west midland meres were variously formed in these ways. Then there are other ways that basins can naturally form, from the collapse of limestone caverns, the damming of streams by blown sand, shingle or landslides, or the work of beavers, and increasingly of man.

There are many times more small ponds than there are moderate sized or large lakes (Fig. 1.5). It is estimated that on a world scale there are close to three hundred million ponds between 0.2 and 1 ha in area, a further twenty-four million between 1 and 10 ha and two million between 10 and 100 ha, which covers the size range most people would think of as a pond or small lake. In contrast the rest of the worlds' lakes number only 20,000 and their combined surface area (including the Caspian Sea with 378,000 km$^2$) of 2.42 million km$^2$ is only a little greater than the 1.82 million km$^2$ of the smaller bodies that average less than 25 ha each. Many of the smaller bodies dot the tundra regions of Canada, Scandinavia and Russia and were formed by soil movements owing to the freezing and thawing of the surface, but a great many are man-made and in the tropics the numbers are probably still increasing. The small irrigation dam, the pool for stock watering, the water supply reservoir and the village fishpond are there the mainstays of modern pond creation.

Ponds probably numbered millions, compared with the perhaps four hundred thousand up to 2 ha in area, present now in Britain and Ireland, before many were filled in to serve the needs of modern agriculture (Fig. 1.6). They once embraced all the functions they currently fulfil in the tropics, and a myriad of other purposes too (Fig. 1.7). There are decoy ponds (for concentrating wild ducks for food in netting pipes that led off them), dew ponds and droving ponds (for watering cattle in the fields or on the move), dye ponds, flax retting ponds (flax is the fibre left after softer tissues have been rotted away under water), blacksmith's forge ponds, hammer ponds (for supplying industrial steam hammers in the nineteenth century), ice ponds (for cutting ice, which was stored in deep vaults, before refrigerators), marl pits (where chalky soil had been dug out for sweetening acid land), mill ponds (to store the water to drive the machinery for grinding grain), moats (for defence of fortified houses), stew ponds (in which fish were kept for the winter when other meat was scarce), swimming ponds, traction engine ponds (steam engines needed lots of water and it had to be replenished frequently), peat cuttings (sometimes quite large in the Netherlands and constituting the Norfolk and Suffolk Broads), and watercress beds (for when watercress and salt constituted a meal in itself). Since the medieval period ponds have increasingly been made for amenity.

Often the functions of working ponds have become redundant but ponds still have a role in conservation and in the appearance of the landscape. They need not

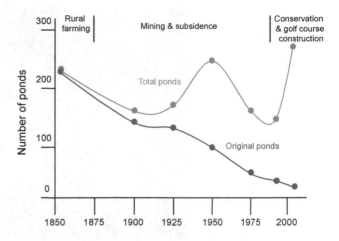

**Fig. 1.6** The number of original farm ponds on the coastal plain of north-east Northumberland has steadily dwindled, but as it did so in the twentieth century, the number of new ponds created from mining subsidence and open cast mining increased. In the late twentieth century, conservation organisations and golf course developers created more ponds. Based on Jeffries (2011).

necessarily be permanent. There are plenty of plants and invertebrates that thrive in temporary waters and some are unusual and of great interest. Though each pond represents only a small part of the landscape, the tendency for them to occur in clusters, linked formerly by only lightly used land, created a distinctive 'patch' ecology, particularly valuable for amphibians. Many ponds, being fed by ground water and thus isolated at the surface, are fishless and amphibian tadpoles do not easily coexist with predatory fish.

The number of ponds present in the 19th century in the UK has now been more than halved and at present rates of loss, there will soon be few left in the agricultural landscape. The reason is partly that they have been filled in because farming practices have become more intensive and ponds get in the way of large machines, whilst cattle grazing, as part of rotations in the use of land on mixed farms, has been replaced by continuous arable cultivation. Partly it is because stock is now watered through pipelines to drinking troughs and the ponds have not been maintained. With time they inevitably silt up. The distribution of newts and frogs in the UK closely echoes the distribution of ponds and their loss is one of the reasons for declining amphibian populations.

Whilst ponds in the agricultural landscape are

**Fig. 1.7** Ponds have been created for a variety of reasons. Top left: Aerial view of the mediaeval fish ponds at St Benet's Abbey, River Bure, Norfolk, used for storing fresh fish over the winter, when the only meat otherwise available was salted. Bottom left: Hale Duck Decoy, Cheshire, built in the 17th century and used until the 1930s for luring ducks, using dogs, for the pot. Top right: Studley Royal Water Gardens, Yorkshire, an 18th century ornamental set of ponds of high formality and little biological interest. Bottom right: A beaver pond in Quebec, no longer common in the UK, but the future may be different. Photographs by Environment Agency, Friends of Pickerings Pasture and Brian Moss.

**rhizome**
underground stem that
grows horizontally

declining, however, the numbers in gardens have been increasing. The grander garden ponds are the shallow mirror lakes designed by the landscape architects of the 18th and 19th centuries to reflect the magnificence of houses often built on the profits of the slave trade, ruthless mining or rabid industry, and from which any wild growth of plants was rapidly removed lest it sully the message of complete control (Fig.1.7). The more modest recent ponds are those of many suburban gardens. Garden ponds are generally very small and have hard edges of paving rather than soft and sedimented edges that will support a plant community. Assiduous owners will grow their plants in submerged pots, and spurn the uncontrolled spread of rhizomes in accumulating leaves and sediment. Goldfish in one or other variety may have been introduced in over-large numbers, and the water is held in with a plastic or rubber liner. The former puddled clay or later concrete linings are too vulnerable to piercing or cracking and modern polymers have taken over. Leaves will be cleared out every winter and plastic nets guard against hungry birds. Nonetheless invertebrates and sometimes amphibians will colonise and provide uncontrolled interest. But it is the rarely managed or neglected garden pond that is of the greatest interest. It will acquire characteristics of its older sisters in the fields and its even older siblings in the tundras and boglands, and among them the world of freshwater ecology, indeed of all ecology, and its significance in the functioning and future of our planet, can be opened up.

# 2 Living in freshwater

## 2.1 A peculiar substance

Whole books have been written about the physics and chemistry of the water molecule. It is very familiar, emerging readily, for some of us, from our taps and cisterns and absolutely crucial to our existence; it is the most precious of commodities in arid lands. It is also a very remarkable and unusual substance. The familiar combination of two hydrogen atoms and one of oxygen belies some deep secrets about how these atoms are linked, on which our entire existence depends.

Life needs to operate in a liquid medium. The insides of cells must be fluid for the biochemical reactions that drive living processes to operate rapidly enough (which would not be the case in a solid medium) and in a controlled way (which the anarchy of a gaseous medium would not permit). The liquid needs also to be at a temperature that is below the flashpoint of carbon compounds, carbon being the most versatile of the elements in forming the huge range of substances that can readily react together and underlie living systems. There are few natural liquids at the Earth's surface or within it. Molten lava, liquid mercury, and carbon dioxide, under high pressure in some crystals, do not qualify for reasons of temperature, toxicity and scarcity, and natural oils are secondary liquids formed by the action of living organisms. Water is not only ideal for the purpose, but also very abundant. It is the only possibility.

**flashpoint**
the lowest temperature at which a liquid gives off enough flammable vapour to ignite

By the usual chemical rules, however, water should be a gas not a liquid. Oxygen is closely related, in a series of increasing atomic weight, to sulphur, selenium and tellurium. Hydrogen telluride, hydrogen selenide and hydrogen sulphide are gases under the temperatures and pressures of the Earth's surface, with lower and lower melting and boiling points along the series. Hydrogen sulphide, at atmospheric pressure, melts at -82ºC and boils at -60ºC. Water (hydrogen oxide) should have even lower melting and boiling points, theoretically about -100ºC and -80ºC, but familiarly melts at 0ºC and boils at 100ºC, which allow it to exist as all of solid ice, liquid water and gaseous vapour at various times and places on the Earth's surface (which has temperatures more than overlapping the 0–100ºC range). The reason for this is that unlike the structure of its somewhat smelly (and very poisonous) fellows in the

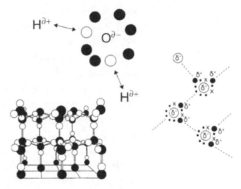

**Fig. 2.1** Oxygen in water attracts electrons shared with its two hydrogen atoms more strongly than do the hydrogens. Each molecule is thus slightly polar and attracts nearby molecules (negative to positive). This gives liquid water a structure, which becomes regular in the crystals of ice (lower left).

series, there is a peculiar linkage between the hydrogen and oxygen atoms in water.

The fellow compounds have covalent bonds. Their atoms share electrons in an even way that leaves them electrically neutral. Water also shares electrons among the hydrogen and oxygen atoms, but the oxygen is slightly more attractive than the hydrogens to them. As a result, the molecules of water are slightly polar, with positive and negative ends. When jumbled together, the individual molecules attract one another and form a sticky framework, most obvious in ice, but still present in the liquid (Fig. 2.1). This attraction means that it takes much more energy to convert ice to water and water to vapour than it does to move hydrogen sulphide, and the other compounds in the series, from solid to liquid to gas. These processes for water can only occur at a much higher temperatures.

There are some further subtleties. Because liquid water retains a lot of the crystal structure of ice, and water vapour retains none of it, it takes much more energy to boil water than to melt ice and much more energy is released when the vapour condenses than when water freezes. You can demonstrate this using a gas or electric cooker. Turn on the heat to a modest level and wait long enough for the supply to be steady, then fill a small pan with ice cubes that have been floating in water for some time to allow them to warm to 0°C, and then drained. Record how long it takes to melt the cubes completely. Continue heating at the same setting until the water is boiling and then record how long it takes to evaporate all the water. The time is a measure of the heat energy delivered. The heat required to melt one gram of water is 79.7 calories and to evaporate it 597.3 calories. The two times should thus be in the ratio of 1:7.5 if you have kept the heat steady. Be careful, of course, with boiling water and hot pans. The consequences for pond organisms are that although freezing readily occurs, so does thawing but more importantly, water does not readily evaporate so that even in dry climates, water persists. Try putting out small, shallow dishes with similar amounts of methylated spirits (which is mostly ethyl alcohol), acetone (nail varnish remover) and water in the open air on a dry day and recording how long it takes to evaporate each. Acetone is not electrically polarised, alcohol is very slightly polar, but not so strongly as water.

Ice floats on water. This is a very unusual property for a solid relative to its liquid, and important for freshwater organisms because it means that in cold weather, ice forms at the surface and then insulates the water below, leaving at least some liquid habitat for fish and invertebrates. It would take intensely cold weather over a very long period to freeze even a shallow pond solid. The reason is again linked to the retention of a crystal structure in liquid water. The molecules are arranged in a well-spaced orderly way in ice. It is a crystal. When it melts, this structure should theoretically completely break down and the molecules should disperse, giving a lower density. But as the ice melts, the molecules remain attracted to one another and collapse inwards, increasing the density of the liquid to slightly greater than that of ice. The dense cold water in a pond then sinks, taking it further away from the surface and insulating it from the place, the surface, where cooling is taking place. You can see this by watching floating ice melt in a clear glass container against a good light. The melt water will show up because of its different refractivity and can be seen as streams falling from the ice cubes. A parallel process occurs when a large building is demolished by explosion. The pile of rubble left occupies a much smaller volume than the original building.

Liquid water is denser than ice up to about 4°C. Further warming then causes the crystal structure to break down progressively so that the liquid becomes less dense as it increases in temperature. This too has consequences for pond organisms. First it takes a lot of energy to overcome the attractions among the water molecules, so that warming water does not change in temperature very quickly, and organisms are not subjected to rapid thermal shocks.

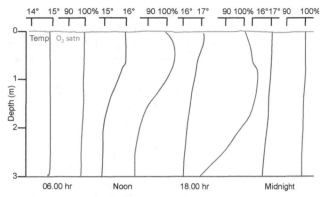

**Fig. 2.2** Typical profiles of temperature and oxygen (% saturation) on a still day in a Somerset pond. The pond temporarily stratifies and oxygen increases towards the surface (because of photosynthesis) but becomes depleted at depth. As the sun sets, however, night-time breezes mix the water back to the more uniform profile it will acquire by dawn.

Secondly, in summer, as the surface warms in the sun, the warm water starts to float on cooler water beneath. This warm surface water then insulates the cooler water below and the temperature layering becomes more marked as the surface heats up more and more. This is called stratification and is a usual phenomenon in lakes deeper than a few metres, persisting for the entire summer. It is prevented by wind mixing in winter, but starts to establish in spring and lasts until surface cooling and rising winds destroy it in late autumn. It occurs in shallow ponds for quite long periods if they are well sheltered by trees, but if they are exposed, the wind is usually enough to keep the water circulating and temperature even from top to bottom. On still, hot days, however, stratification can occur by day and be destroyed overnight (Fig. 2.2). There is much amusement to be had in creating a mini-lake, in a glass mug, by layering very hot tea over very cold milk and finding just how difficult it is, until the tea has cooled considerably, to disturb the stratification by blowing on the surface.

## 2.2 Drag and sinking

Stratification, even on a short-term basis, has many consequences for pond organisms. The deeper water is isolated from the atmosphere and may quickly become deoxygenated by the respiration of bacteria in the water and sediments. Because oxygen is important for all organisms (because it is poisonous for some bacteria and essential for all other organisms) this affects their distribution and growth. Water

*

References given under
authors' names in the
text appear in full in Bibliography and Further
Information on p. 197

does not dissolve much oxygen, but dissolves many other substances very easily and this too is a consequence of its electrically polarised nature. But before I attend to that, there is one last physical consequence of the polarity. Liquid water is sticky. It has a relatively higher viscosity than it would have were not the molecules polar. The attraction between the molecules also lies behind the film of surface tension that forms, and which is strong enough to support the weight of even quite large organisms, like pond skaters, which live and hunt on it. A book in this series (Guthrie, 1989)* deals specifically with these. An organism moving through water is subjected to two kinds of forces: those involved in pushing water out of the way of its progress; and those that drag it back because of the water sticking to its surface. The bigger the surface area in relation to the volume, the more effective are the drag forces, so the smaller the organism, the greater the effect of the drag. (As an object increases in size, its surface area increases by the square of its radius ($r$), but its volume by the cube, so for a sphere, area is $4\pi r^2$ and volume is $(4/3)\pi r^3$, and the ratio of volume to area becomes greater as $r$ increases).

Big organisms like fish are not too inconvenienced by drag, nor are the larger invertebrates, but for many pond organisms, including all those that are microscopic or only just visible to the eye, drag and viscosity are crucial. For them the viscosity effect is much as we would find were we to try swimming in syrup. It is energy-demanding for the movement of limbs that filter particles from the water, but also keeps the particles from bouncing away as the fine hairs of the filters approach them. It helps keep microscopic organisms from sinking (most are denser than water and do not float, despite fond beliefs that they do) too rapidly to the deep (and perhaps deoxygenated) water layers, and this is important for photosynthetic algae because light dwindles rapidly with depth. Shape can be used to counteract viscosity or manipulate it to advantage. Streamlining, where the body is widest about a third back from the head and then tapers, can minimise its effects; for small organisms, flattened or elongate shapes, or the development of spines or star shapes can all enhance its effects and delay sinking.

Stokes' Law, which describes how a body moves in a fluid, and is especially useful in describing sinking rates, can be conveniently tested in the kitchen using a tall glass vase, some Plasticine, play dough or Blu-tack and a dense liquid. You need to scale up the density beyond that of water because plasticine etc. are much denser than are living organisms. Glycerine works well, but alternatively a saturated sugar solution, thick enough to be syrupy, will do and also allows different levels of density depending on the amount of sugar dissolved. Stokes' Law for a spherical body of radius $r$, says that the rate of sinking is a function of size, the difference in density between the body ($\rho_b$) and the fluid ($\rho_f$) and the viscosity, $V$:

$$V = 0.22\ r^2\ (\rho_b\text{-}\rho_f)/V$$

So sinking velocity is proportional to size if you keep the medium and the material of the balls the same. You can vary the size of plasticine balls, from a millimetre or so up to, say, a centimetre, and their shapes, for a given volume. Common shapes are discs, long, thin threads, often slightly curved at the ends, and star shapes with radiating projections or spines. A stopwatch is useful for timing but if the vase is tall enough, the second hand of a watch should suffice.

## 2.3 Dissolution

Water dissolves almost everything to some extent. As a result, and because of the huge range of chemical processes and biological activity that go on in the landscape, a typical natural water will have hundreds of thousands, possibly more than a million different substances dissolved in it. Most will be organic carbon compounds at vanishingly small concentrations, but now detectable because of the sophistication of our analytical machines. Water dissolves most readily those compounds that are polar because it can enter into a reaction with them, so you can dissolve a lot of common salt, washing soda, baking soda, Epsom salts, or nitrate garden fertiliser, all of which are ionic, in it. Some organic compounds, those that contain a hydroxide, carboxy or amino group are also slightly polar and will dissolve readily. They include sugar (sucrose), urea, the amino acids in various pills peddled for health purposes, citric acid in lemon juice or acetic acid in vinegar. In all of these polar compounds, more will react, and therefore more dissolve, up to the point of saturation, as temperature increases.

Non-polar, covalent compounds cannot react with water, and do not truly dissolve. They include oils, fats, hydrocarbons like paraffin, and resins, and also most of the common gases of the atmosphere (except carbon dioxide, which is polar for a similar reason that water is polar). Some of them, the gases and light oils in particular, can enter into

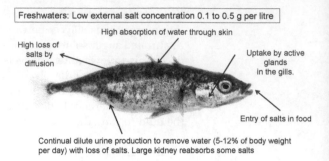

Fig. 2.3 Freshwater animals have problems of absorption of water and loss of salts. They cope with these problems in various ways.

physical mixtures without reaction, if bubbled or shaken with water, but because covalent and polar compounds repel one another, the amount that can so 'dissolve' is very small and the higher the temperature, the less that can stay in the mixture. Water thus dissolves very little oxygen, only about 14 mg per litre at its coldest, and almost none in boiling water, whereas it will accommodate 360 g of common salt, and nearly as much sucrose sugar, at room temperature.

Water chemistry is very important for freshwater organisms. They have body fluids or cell saps that are generally more concentrated than the outside water. (Freshwaters, in their relatively rapid transit through rocks and soils and into lakes and ponds, do not have much time to pick up high concentrations of salts and other substances.) Salts will thus tend to diffuse from within the bodies of organisms into the external water down the concentration gradient, whilst water will tend to diffuse in until the concentrations are equal. The movement of the water is called osmosis. If osmosis is not controlled or accommodated, the result would be that the internal biochemical reactions on which organisms depend could not function in the more dilute internal medium or the cell would burst. Freshwater animals (Fig. 2.3) must therefore continually pump in salts from the water, using energy, through variously located special glands, and conserve them from the food they eat. They must also continually excrete a copious flow of urine to rid themselves of excess water. Freshwater plants, like all plants and most algae and bacteria, have rigid walls of cellulose or other strong polymers. They absorb water until the walls resist further expansion of the cells and osmosis then stops. Plants under water will still lose salts by diffusion, and must continually absorb them through their roots or leaves. In lakes in arid regions, and in the biggest lake of all, the ocean, water flows in but leaves only by evaporation, leaving its salts behind, so that concentrations may be about as high (in the ocean) or much higher (in many arid-zone salt lakes) than in the organisms. The problems then of maintaining salt and water balance are relieved or reversed.

You can demonstrate osmosis easily using dialysis (Visking) tubing, which has properties like those of cell membranes in being permeable to water and salts, but not to big molecules like proteins and starch. A 'cell' can be made by tying off, with cotton thread, a short length of tubing (previously soaked to make it pliable), which has been filled with a sugar or salt solution that has a concentration similar to that of the cell fluids in freshwater animals (say 10 g per litre – kitchen scales and a measuring jug will give sufficient precision). If you leave the 'cell' in a bowl of pond water, it will absorb water and become very swollen and may even burst. Alternatively, if you put it in a concentrated salt solution (seawater has 35 g total salts per litre) the 'cell' will shrink. Or you can take lengths of dandelion scape, pondweed leaves (use the flat-leaved rather than fine-leaved species), pieces of cut potato or carrot, or lettuce leaves, and immerse them in a series of different concentrations from say 0.1 g per litre to 20 g per litre. At the point where the material becomes flaccid, you have a rough measure of the concentration of soluble substances within the cells and can compare it with the concentration in freshwaters (conventionally defined as less than 5 g per litre and usually less than 0.5 g per litre in the UK). At higher concentrations than 5 g per litre, the water is defined as brackish (a term often used in common speech to mean stagnant and distasteful, but it has a precise meaning in science).

**scape**
leafless flower stalk arising from the base of a plant

## 2.4 Oxygen

The second importance of dissolved substances in natural waters lies in the fact that some substances that are required by freshwater organisms are relatively scarce. Organisms require about twenty elements to form their bodies, but even in the most dilute freshwaters, most of these are in ample supply. Two that are naturally very scarce are the compounds of nitrogen and phosphorus and the reasons for this are considered in Chapter 5. It is not that these substances are poorly dissolved, but that they are normally scarce in forms that are available to living organisms on this planet. A third element, oxygen, is scarce simply because of its low solubility. Oxygen is needed for reactions that release energy from food. Some organisms (many bacteria, in fact) can cope without it but the energy release is then very low. For almost all animals and plants, it is absolutely required in the form of the oxygen molecule ($O_2$) and therein lies the problem. Oxygen diffuses into water up to the point of saturation, but the process is slow unless the water is vigorously roiled, and saturation levels are low compared with the demands of active animals. Plants produce oxygen, of course, when they photosynthesise, by effectively breaking down water molecules, but they also use up almost as much oxygen in their own respiration. Oxygen concentrations around plants may rise quite a lot in daytime because of photosynthesis, but they fall equally precipitately at night.

Oxygen can be measured using meters that depend on a plastic membrane covering an electrode. The state of oxidation, effectively the number of free electrons, of the surface of the electrode is converted into a small electric current and registered on the meter. Meters are not inordinately expensive, but not cheap either, and require care if they are to be maintained in reliable working order. If you can afford one, they offer many possibilities for measurements in ponds: for example, how oxygen levels change during a 24-hour period, and how they change with depth in the water or with the season. If you buy a meter, specify that it should have several metres (say 5 to 10) of lead so that you can dangle it into the deeper water (you will need a boat for this, of course). A very useful measure is the percentage saturation of oxygen and most meters will directly measure this. Percentage saturation is the actual concentration divided by the saturation concentration at the particular temperature of the water and most meters will also measure temperature. Most natural waters in summer will be under saturated, except in sunny weather when plants are photosynthesising rapidly and they may become oversaturated for a time. On windy days in winter, they will generally be around saturation because biological activity is then low. Using oxygen measurements over a 24-hour period it is possible (Chapter 7) to work out a great deal of what is happening in the processes of photosynthesis, respiration and decay in the system.

Because they produce it, and therefore have a premium on its availability, plants do not have great problems in obtaining enough oxygen, except in their roots and rhizomes. Fine roots die without oxygen and the sediments where they grow are bacterial fleshpots. Intense decomposition in sediments removes oxygen rapidly and there is limited possibility for its replenishment by diffusion from the overlying water. Diffusion is always slow in a still liquid and even slower in a semi-solid like sediment. The solution that has evolved in plants is to use the faster diffusion that occurs in air (smells rapidly make themselves known), by creating a system of air spaces and tubes within the plant tissues that lead down to similar spaces in the roots, and thence out to the rootlets and the environment close to the root hairs. The tubes are not always continuous within the plant, but the system works well. If you dig into some waterlogged, marshy soil close to a lake or pond, you will usually find that the soil has a grey colour, from reduced iron compounds, flecked with orange pockets around the plant roots, where oxygen diffusing from the roots has oxidised the iron to a rust colour. Cutting across stems and rhizomes of water and marsh plants will demonstrate the air spaces and there might be relationships between the proportion of the stem occupied by air spaces and the oxygenation of the root environment. A surrogate for the latter can be the proportion of organic matter in the soil or sediment, which is measureable by drying some soil, then burning a weighed amount (kitchen scales are now often precise to within 0.1 g), cooling and reweighing to obtain the organic content by difference. Organic content is most conveniently calculated per unit dry weight, rather than wet weight. Soil can be burned in a metal or porcelain crucible on a kitchen stove or hot plate, or with a Bunsen burner, but open the windows because the smell is that of a bonfire, and allow things to cool before handling. The greater the organic content of soil, the lower the oxygen concentrations in the interstitial water are likely to be.

Animals have solved the oxygen problem in different ways (Fig. 2.4). The simplest is by being tiny, up to a few millimetres, so that the surface area is large compared with the bulk, and simple diffusion through the whole surface suffices. Beyond a few millimetres, there are two choices. There can be an expansion of the body surface through gills, as in tadpoles, fish and the larvae of many insects. Many nymphs have flat plates, or bunches of branching filaments, along their abdomens (though sometimes loss of these in mayflies has negligible effects on the survival of the animals). Or a bubble of air, from which oxygen can be absorbed, can be retained under the wing covers or held by hairs on the abdomen. This option means that frequent

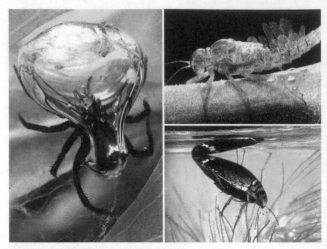

**Fig. 2.4** Oxygen is potentially in short supply in both fresh and saline waters because of its low solubility. Solutions to the problem include storage of air in a silken bell as in the water spider (left), external gills on the abdomen by mayfly nymphs (top right) and entrapment of an air bubble under the wing covers, as in water beetles (bottom right). Photographs by Stephan Hertz and Charles Krebs.

visits to the surface are needed to replenish the bubble. The female water spider lurks in a bell of silk, which she fills with air taken on repeated visits to the surface as she constructs it. Highly motile predators, like some water bugs and beetles, and fish, require higher oxygen concentrations to support their greater activity than sedentary or slow-moving organisms. Over a series of ponds it might be possible to demonstrate an inverse correlation between the numbers or proportion of invertebrate predators and the dawn oxygen concentrations. Oxygen tends to fall to its lowest concentration for the day just before dawn.

## 2.5 Chemical communication

The third important aspect of water chemistry is that the water is the medium of chemical communication for its animals. Just as smells in the air may tell of prey or predators, mates, or the location of water itself for land animals, water also has its dissolved smells that sometimes, in a much cruder way, we humans can detect. For example there is a peaty taste in some brands of Scotch whisky made in the boggy areas of the Western Isles. Swimming animals emit trails of pheromones as they move. Sophisticated microscopic and recording equipment can track the movements of individual copepods as they swim and has

**pheromone**
chemical secreted or excreted by organisms that trigger a response in members of the same species. It may, for example, warn of predators, help locate food or attract mates

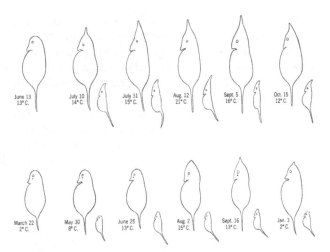

**zooplankter**
animal member of the
plankton

**Fig. 2.5** Many zooplankters undergo changes in shape during the year. These are drawings from a study in the early 1900s showing how two populations of *Daphnia* produce elongated heads (and become thinner) as the summer progresses. The small drawings are of juveniles. *Daphnia* produce a new generation every week or so in summer. The temperature measurements reflect a preoccupation, at the time, that these changes were caused by temperature, but we now know that they are devices to minimise risk of predation by larger invertebrates and by fish.

shown how the trail of a female corkscrewing through the water is closely followed, even when she is out of visual range, by an ardent male. Hitherto we know very little about the substances concerned, except that they exist. Some algae will grow bigger and more complex shapes, which are less easily eaten, when they are grown in water that planktonic filter feeders like *Daphnia* have previously occupied; filter-feeding water fleas (see Chapter 4) will grow odd protruberances (Fig. 2.5) in the presence of invertebrate predators that need to handle their prey in a precise way to capture it successfully, or will become smaller and less easily seen when fish are present. Fish, other than those that filter-feed, must see their prey before deciding to spend energy on an attack. In ponds and lakes with fish, the zooplankters will migrate downwards in daytime to reach darker depths where fish cannot see them (but where their food is scarcer) and upwards at twilight to feed in the dark, food-rich surface waters. But in mountain tarns lacking fish, they will not migrate; migration is costly in energy and selected against when it is not necessary for survival.

The nature of the substance, or substances, emitted by

**copepods**
small, streamlined crustaceans, common, as filter feeders or predators, in the plankton. See Chapter 4

fish that triggers these behaviours has long been sought (see Bronmark & Hansson (2012) for a review of this area). It is likely to be simple and not easily decomposed. Ammonia, trimethylamine and urea have been suggested but do not seem to be responsible. Experiments are possible however even if the substances are not yet identified. Water fleas, copepods and rotifers can often be cultured in jam jars if suitable food is given (see Chapter 4). Experiments can then be made with predators like sticklebacks, dragonfly larvae, water beetles or the larger water fleas such as *Polyphemus*.

# 3 The littoral

**Fig. 3.1** Superficially there seem to be distinct zones of plant communities around many ponds, but the reality is more complex. The eye sees only prominent features and cannot penetrate under water, where the communities change steadily. Bosherston Lily Ponds, Pembrokeshire. Photograph with permission of the National Trust.

'Littoral' means shoreline and is used differently in freshwater and marine ecology. In the latter it generally means the intertidal zone, but freshwater ecologists use it to mean the part of a pond or lake that will support growth of plants or photosynthetic algae on the bottom. Plants are obvious features and a good place to start examining the ecology of ponds. At the edges there will be damp ground, which is not to be ignored because it contains the resting stages of animals that may hatch when water levels rise, and wetland plants that show many of the same features, such as air spaces, as those that grow in the water. Then as the water appears at the ground surface there will be emergent plants, sometimes small rushes or sedges and grasses, sometimes, into deeper water of up to about 1 m, very tall reeds and reedmace, horsetails or sedges. Mixed in with them and then perhaps persisting into water a little deeper, may be floating-leaved plants such as water lilies and free-floating plants, notably the duckweeds in the temperate zone. Completely submerged plants (except for their flowers, which in most cases are borne above the water surface) complete this sequence from land to water. There is a common myth that ponds are surrounded by

**zonation**
the distribution of
animals and plants in
distinct zones

zones of emergent then floating-leaved then submerged plants, and a superficial inspection may seem to support this (Fig. 3.1). Generally this is not true however and you can confirm it by recording presence and absence at close intervals along a rope stretched from dry land into the water. There is usually a zonation of individual species, but a continuous set of mixtures of the various life forms as the water deepens, with solely emergents at the landward edge and solely underwater species in the deeper water, but all sorts of combinations in between.

## 3.1 Plants, birds and amphibians

Aquatic plants are not difficult to identify and the flowers are not often needed. You can use a standard flora (see Chapter 10) if the subtleties of more difficult groups like the sedges, or hybrids and varieties are of interest, or one of the well-illustrated handbooks if the finer details are unimportant to you. As a middle way there is a very easily used, inexpensive and illustrated key to most of the aquatic plants by Haslam, Sinker & Wolseley (1979), available from the admirable Field Studies Council. There is much still to be found out about aquatic plants. The emergents and water lilies tend to be fairly permanent features from year to year, but there tends to be a lot of change in the distribution and species composition of the submerged communities within a pond. Much may depend on local weather conditions, and incidence of grazing by water birds in spring, to determine what the pattern may be in summer.

Submerged plants still retain many of the features of the land plants, from which they relatively recently evolved, such as aerial flowers and pollination by wind or insects. Very few produce flowers under water. On some plants there are the vestigial remains of the stomatal pores, which are used to regulate gas and water exchange on land, but which have no value under water. They can be seen by stripping off some of the epidermis with forceps and examining it under a microscope. I do not think anyone has yet systematically examined many submerged plants for such traces, however. There may be patterns related to their family or life form. Features of land plants tend to increase in the sequence from submerged to floating and floating-leaved to emergent water plants and many emergents produce different sorts of leaves as they grow up through the water column and then above the surface. A comparison of features of the different types in terms of structure, growth rate, vulnerability to grazers and turnover (the rate of replacement of

**vestigial**
forming a remnant of
a structure that has
lost its function during
evolution

leaves, which can be followed by marking individual leaves by tying coloured cotton or wool around their stalks) could be valuable. It is often the case that there have been only limited studies in the literature.

The plant fringes also provide habitat for water birds. Ornithologists tend to concentrate on the rare and exotic but as a result we sometimes have only limited detailed knowledge about common water birds, like coot and moorhen, mute swan, heron, reed warbler and mallard. Long-term observations on distribution and use of the habitat, feeding activity, and changes in numbers might greatly improve our understanding. Bird identification books are readily available. The same is true of amphibians, and ponds are often key habitats for these. There is much information in Trevor Beebee's book in this series that gives clues to possible investigations (see Chapter 10).

Methods for sampling and observing the larger organisms, like birds and plants, need no special mention. Binoculars and a hand lens are respectively useful for proper identification. If a pond is too deep for wading, a grapnel, made by lashing two garden rake heads back to back and tying them to a rope, can be used to drag for a sample of the plants. On a large pond or small lake, the material washed up at the edge often gives a sample of the most abundant species. Plants pose few difficulties, except that there is one important group of large green algae, the charophytes, or stoneworts, which have a distinctive structure (Fig. 3.2) and are not included in the standard

**Fig. 3.2** Stoneworts are large green algae that have a central axis made up of large single cells, alternating with plates (nodes) of smaller cells from which whorls of similarly structured branches emerge. They are anchored in sediment by colourless filaments called rhizoids.

**whorl**
a ring of plant parts radiating out from the same point on a stem

**hybridise**
form hybrids by inter-breeding between species

floras. Once they are recognised by their often distinctive smell (garlicky), whorled branches, lack of flowers and often a rough feel engendered by deposits of calcium carbonate (hence stonewort) on their surfaces, a specific flora (John *et al.* (2011) is now the standard work) will readily give a name. There are some small difficulties because species tend to be quite variable and some hybridise.

## 3.2 Light penetration and the littoral zone

Light is rapidly absorbed by water, with different wavelengths absorbed at different rates, but none of them penetrating very deeply except in those ocean waters that are so remote from land and nutrient-poor that very little grows or is suspended in them. Freshwaters accumulate dissolved coloured substances from the water washed into them from the soils of the land, and from the decay of water plants and algae. They are more nutrient-rich than the deep ocean and have much greater populations of algae, and particles of organic debris or eroded soil suspended in the water. Although the most penetrative violet and blue light may reach a couple of hundred metres in the ocean, and a few tens of metres in a large, clear mountain lake, it is unusual for any light to be left below a metre or two in a pond. The land has a very strong influence. Red light is the least energetic and in pure water penetrates least, and the shorter the wavelength, the more penetrative, but in pond waters, the yellow colour of washed-in organic matter takes out blue and violet light very readily, so that the most penetrative wave bands tend to be in the green and yellow. The wave bands that are least readily absorbed thus give many pond waters their often murky green or brown colour, just as deep ocean water appears blue.

The light is absorbed exponentially (Fig. 3.3), which means that for each successive depth interval, in a well mixed pond, a steady proportion of the light is absorbed, until an indetectable amount is left. The key depth, however, is that at which there is insufficient light left for an alga or plant on the bottom to be able to photosynthesise enough to meet its own maintenance needs, and therefore to grow. This is called the euphotic depth and defines the point at which the littoral zone of the pond or lake ends. Many ponds are so shallow that the entire bottom may be littoral, but others may have some of their bottom in the profundal zone, below the euphotic depth. Light meters are expensive, but a useful approximation can be made with a Secchi disc, which is a weighted metal disc, the size of a tea or dinner plate,

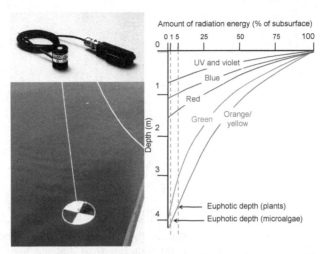

**Fig. 3.3** Light penetration is best measured with a radiometer (top left) able to discriminate different wavebands. In pure water, the shortest, most energetic, wavelengths (UV, violet, blue) are most penetrative, and the longest (red, infra red) most rapidly absorbed. But natural waters are stained with yellow-brown organic compounds, which absorb readily at the blue/violet end of the spectrum. They are also loaded with suspended matter that readily absorbs these wavebands. In consequence green and orange-yellow light penetrate deepest, but at best not nearly as deeply as light would penetrate in pure water. Most of the radiation is taken out within a few metres. The levels at which 5% and 1% of the immediately sub-surface light of the most penetrative waveband are conventionally taken as the depths (euphotic depths) at which the respiration of plants and microalgae respectively are just compensated for by the amount of photosynthesis possible. Immediately sub-surface depth is taken as the reference rather than the surface because a great deal of incident light is reflected at the surface, and is not available for absorption. A very approximate estimate of euphotic depth is given by dangling a weighted Secchi disc (lower left) to the depth at which it just disappears to the eye. This depth is about 40% of the euphotic depth.

held by wire or cord, or mounted at the end of a pole, so that it hangs flat in the water from a rope or the pole by which it can be lowered. Admiral Secchi, in the 18th century, indeed did investigate the transparency of the Mediterranean Sea by dangling one of his ship's dinner plates. Modern discs are painted in alternating black and white quadrants. The mean of the depths at which the disc just disappears on lowering and reappears on raising is taken as the Secchi transparency and is usually about 40% of the euphotic depth, which in turn is about 1% of the light measured just under the surface of the water. Secchi depths are usually taken from a boat or from the edge if there is a vertical bank

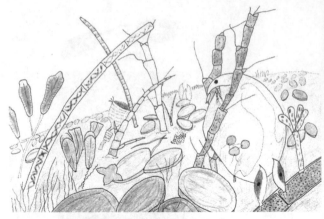

**Fig. 3.4** The landscape of a leaf (simplified, but not unrealistic). A large cladoceran, *Eurycercus lamellata*, that feeds on periphyton pauses among filaments of *Bulbochaete*, whilst other filaments of *Ulothrix*, *Stigeoclonium* and *Spirogyra* (to the left with spiral chloroplasts) and stalked diatoms (*Gomphonema*, left) contribute to a forest that covers the leaf. Attached to the leaf surface by mucilage-producing organs called raphes, the oval diatoms, *Cocconeis*, are largely immune to grazers, unless the whole leaf is consumed. There are bacteria and a ciliate, *Vorticella*, which feeds on bacteria and very small algae, waits with its anchoring filament coiled.

**pelagic**

originated as meaning the deep ocean, with its sharks and whales, away from the intertidal zone but, as often happens when scientists want to sound important, its meaning has been modified in freshwaters to mean the offshore part of a pond or lake, with water deeper than the euphotic depth

**profundal**

the water below the euphotic depth and the sediment underlying it

into deep water, but can be taken by wading if the bottom is not unduly stirred up. Microalgae will grow on the bottom at lower light intensities than bulky plants, because their chlorophyll is not masked by plant tissues, so the euphotic depth for plants is generally shallower than for algae, 5% rather than 1% of the subsurface light, in very general terms.

The littoral zone is the richest in biodiversity compared with the darkened mud of the profundal and the less structured plankton community (Chapter 4). There is a limited plankton among the plants but it becomes more characteristic in the offshore pelagic zone as the plants dwindle away. The plankton communities within the littoral and in the pelagic share many species, but each has its specialists. Among the plants are many semi-planktonic organisms that may attach themselves to the plants, but also detach and move or be suspended in the water between them; in the pelagic zone, organisms either swim or are maintained in suspension by eddy currents generated by the wind. At the bases of the plant beds, the organisms of the sediments have much in common with those in the more limited community of a profundal zone, so I have given no separate treatment to this. Macroscopic algae, or masses of microscopic gelatinous or filamentous algae also provide some of the structure of the littoral zone.

We are thus faced with quite a complex system in which the plants, amphibians and birds are easily identifiable, (and in which fish are readily identified but difficult to study), but in which major roles are played by microorganisms, particularly bacteria, microalgae and invertebrates, which dominate the biodiversity (Fig. 3.4). Alas, the ponds of the UK no longer are graced by the large mammals that were once crucial in moving nutrients, through their grazing and dunging, between the ponds and wetlands and the drier land. But were they still with us, beavers, hippopotami, elephants and mammoths, aurochs, wild horses and bison would pose us no more identification problems than the otter, water vole, water shrew and introduced mink that have managed to hang on.

## 3.3 Fish

Fish, however, are very much with us and for many people, fish epitomise freshwater ecology (Fig. 3.5). The British fish fauna is limited, because Britain was isolated by rising sea levels soon after the ice of the last phase of the recent Ice Age melted back, and only a few species had time to return through rivers that previously connected Britain to the mainland. However, the forty or so native species have been boosted by about as many alien species that have been introduced for various purposes or accidentally. They are

**Fig. 3.5** Perhaps the commonest pond fish in the UK is the three-spined stickleback (*Gasterosteus aculeatus*). It has a complex life history in which males become brightly coloured before going through a behavioural ritual, mating, and construction of a nest of plants or filamentous algae, in which the male cares for the young.

nonetheless only parts of a complex system and no more (or less) important than anything else. Small fish, such as fish larvae and sticklebacks, can be dipped out (if you are quiet and quick) of small ponds with a long-handled pond net, such as is used for invertebrates (see below). For larger ponds and small lakes, there are several methods: rod and line, gill nets, fish traps, electrofishing, and seine nets, some of which are more ethically acceptable than others and for all of which a licence from the national environmental agency is usually required. The ideal method for serious study is unselective of size and species and does not damage the fish; but the ideal method does not exist.

Gill nets, set at night, floating vertically in the water, entangle the fish by their gills and kill them but give a representative sample if a range of mesh sizes is used, and are the preferred method for studies of population size and changes. They can be used in the open water or close to plants beds to give samples of all but those species that remain close to the bottom. Seine nets are paid out from the shore in an arc and then brought back to shore. They have floats at the top and weights at the bottom and their successful deployment requires quietness and least disturbance in setting, and care in ensuring that the weighted bottom is never lifted. If it is, the fish will escape underneath. The net is dragged slowly in by at least two people at either end and will generally enclose fish, though the community will have been selectively sampled. Large fish that swim fast will generally escape and the larger sizes of fish that move out into the open water away from the shore will be missed. The sample will generally be of juvenile fish. Fish traps include fyke nets in which curious fish enter a long netting tunnel, periodically constricted, with a dead-end and have difficulty in finding their way out. Barriers need to be placed at the entrance to deter otters and hence the nets are selective for smaller fish and attract bottom feeders like eels rather than open water feeders. Electrofishing is a specialised business in which an electric field, powered by a bankside generator, is created between electrodes, one of which has a net attached. Fish are attracted to this electrode and can be pulled out of the water stunned but alive. The method works best in waters of low conductivity and can be hazardous for the operators. Proper training and licensing is required. Rod and line is highly selective and entire books have been written on how to lure particular species.

Once caught, fish should be kept in a large tank of water. They are easily damaged and deoxygenated. Ideally air or oxygen is bubbled into the tank and the fish should be released as soon as possible after identification and measurement for which it may be necessary to anaesthetise them using a few drops of clove oil. They will recover when put back into fresh water. There are good identification guides and the only real problems are with fry and very young fish, and sometimes with hybrids such as those of roach and bream. For population studies, the length from snout to the fork of the tail is normally measured, together with weight and condition. Fish can be aged using the rings on scales, usually taken from the region just below the dorsal fin in the middle of the back. Plastic forceps are used to remove a few scales to a small paper envelope so that they can be examined under a microscope. The rings are formed because growth is checked by low temperature and reduced food supply in winter. In farmed fish, fed year around, there may be no rings or irregular ones. Sometimes a few fish are killed so that they can be dissected and their gut contents examined. The only ethical way of doing this is by giving an overdose of anaesthetic. However, although fish are crucial to the understanding of how pond ecosystems function, the better way of studying them for amateurs is to use small live fish in small-scale tank experiments. Fish communities in larger ponds and lakes are usually best left to professional organisations with large-scale facilities and sufficient staff to ensure the fish are well treated.

**fry**
recently hatched fish

## 3.4 Microorganisms

Microorganisms and invertebrates, however, are ideal for amateur study. There are two ways of considering them. The first is by function (their traits and what they do) and the second is by name. Ideally we need some degree of identification before we can say much about function, though that is not often possible with bacteria. We need also to be able to sample them for inspection.

The bacteria are particularly problematic. We used to think that there were rather few of them and that we could culture them on agar plates in the manner that disease bacteria are cultured in medical laboratories. It turned out that almost all of the greatly varied bacteria that could be seen with advanced microscopy could not easily be cultured, and very few have been. They come in a variety of shapes (Fig. 3.6) and sizes beyond the familiar rods and cocci of medicine and it is very difficult to define a species because they continually swap genes as they intermingle in the sediments or on surfaces or suspended in the water.

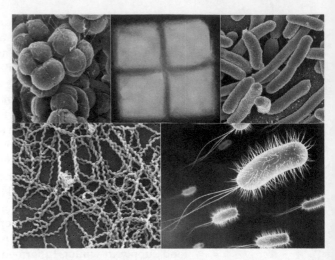

**Fig. 3.6** Freshwater bacteria come in a variety of shapes and mostly have never been cultured in the laboratory. A great many are known only from differences in their nucleic acid (DNA and RNA) base sequences. Archaea and bacteria are structurally the simplest of organisms, but nonetheless in reality are highly intricate. Upper left and centre are two examples of Archaea, the first from very hot water, the second from very salty water, but Archaea are known to be abundant in less extreme habitats also. Eubacteria include rod-shaped cells, upper right, which are about 1 μm long (a micrometre is a thousandth of a millimetre) and a spirochaete bacterium (lower left), with cells 20 μm long but only 0.5 μm thick. More sophisticated techniques reveal more details as in the common gut bacterium, *Escherichia coli*, which swims with fine hairs, or flagella (lower right). From Moss (2012).

They are now characterised by sequencing their nucleic acids in mixed samples from the water and this requires expensive techniques and equipment and tells us relatively little beyond presence or absence. Nonetheless in terms of the versatility of the chemical processes they carry out, the bacteria are the key organisms for the functioning of lakes and ponds, indeed of the whole planet.

It is they that fix nitrogen (convert nitrogen gas to amino groups) from the atmosphere and return it through processes such as ammonification (amino groups in proteins broken down to ammonia), nitrification (ammonia to nitrate) and denitrification (nitrate to nitrous oxide or nitrogen). It is they that cycle sulphur through hydrogen sulphide, sulphate and a variety of other, often smelly, sulphur compounds, and that create the conditions for solubilising iron and manganese, both absolutely required by all organisms. They also degrade organic matter to methane, an important greenhouse gas, in deoxygenated sediments.

And it is they that power the bulk of the cycles of carbon and oxygen through their immense versatility in decomposing millions of different carbon compounds, and which have, for billions of years, controlled the composition of the atmosphere and maintained the planetary surface within a range of temperatures in which liquid water can exist. There was a long period, nearly 2 billion of the 4.54 billion years (on current estimates) of the planet's existence, when only bacteria were present and they have been continually present for at least 3.8 billion years. A biosphere without any plants, fungi or animals is conceivable; one without bacteria is not. Though determination of the precise nature of the bacteria is beyond the reach of amateur laboratories, their activity, especially in decomposition and photosynthesis, is not. One particular group, the cyanobacteria or blue-green bacteria, formerly the blue-green algae, is very important and more readily identifiable than the others and amenable to the techniques used for the algae.

## 3.5 Algae, protozoa and microfungi

'The algae', however, is no longer seen as a single coherent group. With the Protozoa and the aquatic microfungi, their classification has been repeatedly overturned as new information has become available on their fine structure from the electron microscope, and on their genes from studies of their nucleic acids. It will help to recount a little of the progress of evolution to understand why the situation is now much more complicated than it was. Before the 1960s we recognised two kingdoms: Plants (including fungi), studied in botany departments and Animals, in zoology departments. Bacteria were the province largely of medical departments. By the late 1960s a greater sophistication had entered and we used a simple classification of kingdoms (Key A).

## Key A Traditional key to kingdoms of organisms

**1a** Cells without internal membrane-bound organelles. DNA not borne on protein bodies, but existing as circular plasmids suspended in the cell matrix
**Prokaryota**

**1b** Cells with internal membrane-bound organelles, including a membrane-bound nucleus containing DNA born on protein bodies (chromosomes)
**Eukaryota, 2**

**2a** Organisms specialising on a single type of energy source (feeding mode: photosynthesis, food in solution, solid food), unicellular or multicellular with tissue differentiation
**3**

**2b** Organisms usually having mixed energy sources, or clearly structurally related to species that are mixotrophic (using more than one feeding mode); unicellular or colonial but without tissue differentiation **Protista**

**3a** Cells with membranes but not rigid walls, unicellular or multicellular, feeding by ingestion of particulate food (phagotrophy) **Animalia**

**3b** Cells with rigid external walls, obtaining energy by photosynthesis (autotrophy) **Plantae**

**3c** Cells with rigid external walls, obtaining energy by absorbing soluble compounds following action of enzymes secreted into the food source (saprotrophy)
**Fungi**

**symbiosis (plural: symbioses)**
the living together of two different species

In that system, the larger algae (the seaweeds and charophytes) were placed with the land plants, and most of the microalgae, together with aquatic fungi and single-celled animals, the protozoans, were placed in the Protista. The concept was that eukaryotes had evolved from prokaryotes through a set of symbioses that had brought together saprotrophic and photosynthetic bacteria, with a host prokaryote that was capable of engulfing particles and had ingested other, smaller bacteria. These became the internal organelles of the eukaryotic cells. Such symbioses had been frequent and widespread on the ocean floor and may have occurred in response to a rise in oxygen concentrations brought about by the proliferation of the cyanobacteria, which had evolved oxygen-releasing photosynthesis. Oxygen was toxic to bacteria that had previously not been exposed to it. The outcome was a very complicated and varied set of single-celled eukaryotes, mostly quite

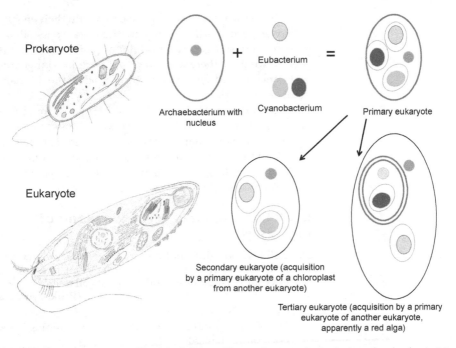

Prokaryote

Archaebacterium with nucleus

+ Eubacterium = 

Cyanobacterium

Primary eukaryote

Eukaryote

Secondary eukaryote (acquisition by a primary eukaryote of a chloroplast from another eukaryote)

Tertiary eukaryote (acquisition by a primary eukaryote of another eukaryote, apparently a red alga)

**Fig. 3.7** Cells are either prokaryotes (eubacteria and archaebacteria) or eukaryotes. Top left is a drawing of a eubacterium and bottom left of a eukaryote. Cells will not necessarily have all of the sub-structures shown; these are composite drawings showing all possibilities, but prokaryotes will always have an external membrane and some DNA plasmids (the thin loops). Eukaryotes will always have an external membrane, a nucleus (shown in pale blue) containing chromosomes with DNA, and some mitochondria (grey structures with internal folds). Current views on how eukaryotes became assembled from prokaryotes are shown on the right. An archaebacterium, presumed to contain some sort of nucleus, acquired eubacteria, which became mitochondria and sometimes cyanobacteria, which became chloroplasts. Cyanobacteria have both blue-green and reddish pigments, with one or other predominating. The primary eukaryotes may have been animal-like with only mitochondria or green algal or plant-like with greenish chloroplasts, or like red algae, with reddish chloroplasts. Acquisition meant that the mitochondria and chloroplasts acquired a second membrane where the bacteria had been enfolded by the host's membrane. Secondary eukaryotes (for example euglenoids) were formed when a primary eukaryote acquired chloroplasts from another eukaryote so that the chloroplasts have three membranes. In tertiary eukaryotes (such as diatoms, yellow algae, cryptomonads and dinoflagellates), an entire eukaryote, a red alga, was acquired. Sometimes only the chloroplast of this red alga has been retained, with four membranes, sometimes more of the structure is apparent, and sometimes only a few genes are left to tell us of this past event. There are many variants on the themes of acquisition and then loss of component parts.

versatile in the ways they were able to obtain their energy. As oxygen levels further increased, there were possibilities for evolution of larger, multicellular organisms and as this process went on, there was a specialisation into kingdoms that concentrated on single feeding modes, the animals, plants and fungi.

That outline is still broadly valid (Fig. 3.7) insofar as symbiosis is concerned though we know now of several layers of symbiosis, but the rest has changed. The current view (and there is no guarantee that it will not continue to change) is simplified in Key B. It illustrates the distinctions but is not intended for identification.

## Key B Contemporary key to kingdoms of organisms

**1a** Cells without internal membrane-bound organelles. DNA not borne on protein bodies, but existing as circular plasmids suspended in the cell matrix
**(Prokaryota), 2**

**1b** Cells with internal membrane-bound organelles, including a membrane-bound nucleus containing DNA born on protein bodies (chromosomes)
**(Eukaryota), 3**

**2a** Membranes with glycerol ether lipids     **Archaea**
**2b** Membranes with glycerol ester lipids     **Eubacteria**

**3a** Cells formed by symbiotic eubacteria invading an archaean host     **4**
**3b** Cells formed by invasion of a eukaryotic cell by organelles, usually a photosynthetic plastid or a mitochondrion derived from another eukaryote. Distinctive flagella with complex structure of fine hairs, born at the front of the cell and a lateral groove
**Excavata**
(euglenoids, trypanosomes and several other groups of colourless flagellates formerly classed as Protozoa)

**4a** Cells formed by invasion of a eukaryotic cell by at least one other eukaryotic cell     **5**
**4b** Cells photosynthetic owing to symbiosis with a cyanobacterium, flagella at front of cell where present     **Archaeplastida**
(glaucophytes, red algae, green algae, plants)

**5a** Cells not photosynthetic, single flagellum at the back of the cell propelling cells forward
**Opisthokonta** (animals and fungi)

**flagellum (plural flagella)** long whip-like appendage used for swimming

**5b** Not like this, flagella when present at front of cell, pulling cells through the water **6**

**6a** Cells lacking flagella and rigid walls, amoeboid
**Amoebozoa** (amoebas, slime moulds)
**6b** Not like this **7**

**7a** Cells often amoeboid and with supporting external or internal mineral structures of calcium carbonate or silica
**Rhizaria**
(former Protozoa like the radiozoans, foraminiferans and vampyrellids)
**7b** Not like this **8**

**8a** Cells formed by invasion of a eukaryote cell by a red alga **9**
**8b** Cells formed by invasion of a eukaryote by a red alga and then subsequently by further eukaryotes
**Alveolata**
(including dinoflagellates, ciliates and several other former protozoan groups)

**9a** Flagella without fine hairs, mixed feeding modes
**Hacrobia**
(haptophytes and cryptophytes from former algal groups, katablepharids, telonemids and others from the former Protozoa)
**9b** Flagella with distinctive fine structure with complex fine hairs, mixed feeding modes **Harosa**
(former algae of the xanthophytes, brown seaweeds, chrysophytes, diatoms; bioecids and opalinids from the former Protozoa; and oomycetes and labyrinthulomycetes from the former fungi)

**anaerobic**
living where oxygen is not present

The primary distinction of prokaryotes and eukaryotes remains but the former are now represented by two kingdoms, the Archaea, which are all anaerobic forms, often living in extreme environments, like hot springs or deep sea thermal vents or very acid rock seepages, and the Eubacteria which are the more familiar ones in soils, waters and as parasites and disease organisms. Eukaryotes have a nucleus containing DNA on complex chromosomes surrounded by a nuclear membrane and other membrane-bound organelles. Some eukaryote kingdoms originated as a simple symbiosis of several bacteria, including probably an archaean that became the cell envelope, others that became the mitochondria, responsible for many energy exchanges within the cells, and yet others that were photosynthetic cyanobacteria that became the chloroplasts, where these are present. There are only three groups of algae that are so simple, the glaucophytes, green algae and the red algae and they are classified now with the plants, and are placed in a kingdom called the Archaeplastida. When their cells have flagella they arise at the front of the cell and row it through the water. The animals and fungi are also closely related in the kingdom Opisthokonta and are simple eukaryotes but when they have flagella, they arise at the rear of the cell and propel the cell forwards. You may have seen videos of moving spermatozoa that show this well. The surprising juxtaposition here is the link between animals and the apparently very different looking fungi, but they share a large number of gene sequences. One group of former animals, the Amoebozoa turns out to have rather different genes than the rest of the animals and is placed now in a separate kingdom.

All the other kingdoms are of complex eukaryotes in which there have been multiple symbioses, where a second and even third eukaryote has invaded an original simple eukaryote host. Often a red alga has invaded or just the chloroplasts of another eukaryote cell. Sometimes the electron microscope reveals very obvious cells within cells in these cases, but there have often been losses of parts or all of the new symbionts, so that the evolution of the cells can only be traced through their DNA, which retains remnant genes of this complex ancestry. Thus it is that the dinoflagellates, very distinctive cells with two flagella, one borne in a groove around the middle of the cell and a cell covering of cellulose plates, have an ancestry involving symbiosis of a green alga by a red alga (there are microscopic red algae; they are not all large seaweeds) and then subsequently in some cases by a third symbiont, a diatom, another green alga or a cyanobacterium. Even more surprising, the protozoan causing malaria, now an apparently simple blood parasite, has a related ancestry but has lost all of the internal structures that indicate the successive symbioses, but has retained a number of tell-tale genes specific to red algae. Key B details the relationships showing how organisms are now classified into nine kingdoms and explains the fundamental differences, but it is of little help in practical identification. Keys and a strategy are given below which should overcome this problem.

## 3.6 Sampling and identification of microorganisms

A compound microscope, ideally with water or oil immersion lenses, is essential if you are to look at the microorganisms, and if you have one, studies of ponds reach a completely new dimension. Putting a drop of water on a slide may not reveal very much but if the organisms are concentrated in a centrifuge or through a net, you will find plenty of interest, and this is considered in the next chapter on the plankton. Electrical laboratory centrifuges are expensive, but hand operated ones are not. You can even make one; instructions are given in John *et al.* (2011) but buying the book will cost you more than buying a hand centrifuge! If you scrape the film of organisms with a knife or stiff brush from any underwater surface (rocks, plants, dead wood, old bottles), suspend it in water and examine a drop of it, you will see a great diversity, including bacteria, algae and protozoa. You can also place artificial surfaces (glass slides in suitable racks, ceramic tiles, plastic pan scrapers, plastic sponges) in the water for a few weeks to grow such communities. Scrape the hard ones, squeeze out the scrapers or sponges. There is still controversy about whether such convenient artificial substrates grow the same communities as natural ones and there is scope for investigations of this. There is also uncertainty as to whether different species of aquatic plants have specific communities determined by the nature of their surfaces.

Mud (Fig. 3.8) can be sampled using a long glass or plastic tube, about a metre long or a little more. The tube is held at a low angle above the horizontal with your thumb blocking the top end and the lower end at the mud surface. You draw the bottom of the tube across the mud surface as you release your thumb and a mixture of water and surface sediment will be drawn up into it. Run the mixture into a jar and repeat the operation until you have a couple of centimetres or so of settled sediment. Allow settling to be completed over a few hours in the dark and then gently pour or siphon off as much as possible of the water. Swirl the sediment to mix it then pour it into a small dish to give an even flat layer. If you place microscope cover slips, or small pieces (about 1 cm$^2$) of lens tissue or fine, open, thin silk fabric on the surface, and leave in the light for several hours, algae and protozoa will migrate to the surface and adhere to the cover slip (which can be removed, mounted in a drop of water and examined directly under the microscope) or entangle in the lens tissue or silk, from which they can be

**Fig. 3.8** The terrain of the sediment. Sand grains, like small hills project from finer organic material and clay. On them diatoms and other algae seek the light. Most algae on sediments can move for there is always a risk of burial and a need to return to the surface. A large filament of the cyanobacterium, *Oscillatoria* slides over the grains from the right whilst near it, a large *Amoeba* extends its pseudopodia to engulf bacteria and small diatoms. Two flagellated cryptomonads hover over the sediment in front of a pea mussel, *Pisidium*, which pumps in water and disturbed sediment through one of its siphons, filters out edible material then pumps out the cleared water. Oligochaete worms (*Tubifex*) burrow head downwards into the underlying more organic material, whilst a cell of *Phacus* (left) takes advantage of the ammonium supply that diffuses from the sediments where the bacteria are decomposing the organic matter and providing the ultimate food for the worms. The jaggedness of the mineral grains tells us that this sediment is receiving eroded soil. Grains that have been rolled around for a long time in the water and abraded each other tend to be much smoother.

shaken into a very small volume of water. There are possibilities for investigation here concerning whether there is a daily rhythm of movement between light and dark and whether the cover slips (which may become deoxygenated underneath) select a different community from the more permeable tissue or silk. You now have samples to examine and the problem becomes one of identification.

There will be a large range of organisms in most samples, sometimes an overwhelming diversity, and some will not be easily identifiable. The sensible approach is to tackle the larger and more distinctive microorganisms first and to examine them live. They can be preserved with a few drops of iodine solution (iodine dissolved in potassium iodide solution and available from pharmacists; concentrations vary but add enough to give a brown solution the colour of strong tea) but you will learn much more from live specimens. Some organisms will be unidentifiable because

**motile**
able to move

they are resting stages, spores, fragments or just too small. This latter applies especially to bacteria and colourless flagellates. Even professionals have that problem, but a lot can be done with what is identifiable.

Key C is a pragmatic one that will allow you to recognise the main groups of visually distinctive microorganisms. It will help you find a more specific key (given later or listed at the end of the book) that in many cases will give you a reliable name, if only to genus. Some multicellular animals, like nematodes, may overlap in size range with the protozoans, but can be distinguished by a rapid whip like movement, pointed ends and internal organs. I have given a variety of names, used by different people, which should help you find the right keys and also link in with the modern classification in Key B.

**cottage loaves**
a cottage loaf is charac-
terised by its shape - two
round loaves, baked
with one on top of the
other, with the upper
loaf being somewhat
smaller

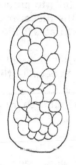

C.1

## Key C Pragmatic key to groups of microorganisms

| 1a | Not pigmented, grey or colourless | 2 |
| 1b | Distinctly coloured | 9 |

2a Non-motile, rounded, or, if filamentous, moving very slowly, sometimes almost imperceptibly or if attached to a substratum not by stalks, and not moving at all **3**

2b Motile, sometimes jerkily, sometimes more steadily, or attached by stalks to a substratum often with obviously beating cilia **6**

**cilium (plural cilia)**
hair-like structure that
projects from the cell
surface and can move

3a Very small (<2 µm in at least one dimension), colourless, rounded, lozenge shaped, or very fine filaments; may move very slowly but not jerkily
**Non-photosynthetic bacteria of many kinds**

3b Larger than 2 µm in all dimensions **4**

4a Large grey granular cells (C.1), ovoid, sometimes like cottage loaves when dividing and containing large granules of sulphur
*Achromatium* (sulphur bacterium)

4b Filamentous **5**

5a Very thin with dark granules irregularly placed along the filament (C.2), usually growing on sediments
*Beggiatoa* (sulphur bacterium)

5b Whitish filaments emerging from carcasses of fish
*Saprolegnia* (once a fungus, now Harosa)

C.2

C.3

**6a** Very small (up to 10 μm), jerkily motile, often very rapidly **Colourless flagellates** (heterotrophic nanoflagellates)

**6b** Larger, slow moving     **7**

**7a** >10 μm up to 200 μm, moving slowly by amoeboid movement (in which the cell extends projections and then the contents flow into them); may have a 'shell' or testa     **Amoebozoans**

**7b** Moving by flagella or cilia but not amoeboid     **8**

**8a** >10 μm up to 200 μm, motile, but steadily not jerkily, by a flagellum emerging from a gullet or indentation at one end (C.3), sometimes with a helical appearance and sometimes able to change shape **Colourless excavates** (euglenoids, Euglenophyta)

**8b** Larger (>10 μm up to 1 mm), moving rapidly with a smooth gliding motion, or attached by stalks to surfaces, sometimes in colonies, usually with the flickering of beating cilia evident     **Ciliates** (Ciliata)

**9a** Grass green (but older cells may be paler and yellow, but will stain black with iodine solution because they contain starch)     **10**

**9b** Not grass green or staining black with iodine solution     **13**

C.4

**10a** Motile with one or two flagella (unless resting!)     **11**

**10b** Not motile by flagella. There may be a slow creep     **12**

**11a** Single cell or colonies, actively moving. Two flagella may be visible and often a red eyespot close to them **Green algae** (Chlorophyta, Volvocales)

**11b** Single cell, actively moving, usually with one visible flagellum, and an eyespot; helical organisation to cell sometimes apparent (C.4), frequently in organically rich habitats **Excavatae** (Euglenophyta, green euglenoids)

**12a** Single cells, filaments or colonies, not obviously motile, without a distinct constriction dividing the cell into two halves **Green algae** (Chlorophyta, several groups)

**12b** Single cells or filaments, with each cell divided by a constriction across the middle, sometimes shallow, sometimes very deep. Richly diverse in acid waters, but three genera (*Cosmarium, Closterium, Staurastrum*) common in alkaline waters **Green algae** (Chlorophyta, Desmidiales, desmids)

**13a** Blue, greyish-blue or reddish               **14**
**13b** Yellow, orange or brown                **20**

**14a** Cells blue, blue green, sometimes pinkish or, if old, can be yellowish         **15**
**14b** Cells distinctly red         **17**

**15a** Pigments in an internal chloroplast so that parts of the cell are coloured and other parts are grey or colourless (C.5 left). Single cells. Pigments distinctly blue or blue-green     **Glaucophytes** (Glaucophyta: probably former green algae that have lost their chloroplasts but then been recolonised by cyanobacteria)

**15b** Uniformly coloured throughout (C.5 right). This is one of the most difficult decisions. If you can see distinct non-pigmented areas, take 15a, but sometimes cells that do have distinct chloroplasts may appear coloured throughout. Look at several cells and look for fine lines where the separate chloroplasts abut     **16**

**C.5**

**16a** Cells all of one kind single, filamentous or colonial (and often surrounded by mucilage, which can be shown up by mounting with a tiny drop of Indian ink added) uniformly coloured throughout (pigments not in an internal chloroplast so that parts of the cell are grey and other parts coloured)
    **Non nitrogen–fixing Cyanobacteria** (blue-green bacteria, formerly blue-green algae)

**16b** Filaments with two sorts of cells, mostly uniformly coloured throughout but also with cells (heterocysts) that are larger, colourless, and at the ends of, or spaced along the filament     **Nitrogen–fixing Cyanobacteria**

**17a** Single cells, but usually in groups; distinctly red, usually on a substratum, not motile, without flagella
    *Porphyridium* (red algae, Rhodophyta)

**17b** Cells motile and microscopic, or non-motile and macroscopic     **18**

**18a** Cell motile and microscopic; reddish (sometimes blue) with a tinge of brown, motile by flagella and not attached to surfaces     **Cryptomonads** (Cryptophyta, Hacrobia)

**18b** Macroscopic, attached to rocks or occasionally plants     **19**

**19a** Appear as red covering on rocks, like splashes of red or magenta paint     *Hildenbrandia* (red algae, Rhodophyta)

**19b** Filaments that en masse look grey-blue or grey-yellow, most common in streams but sometimes in rocky ponds **Red algae** (Rhodophyta)

**20a** Generally rich brown colour **21**

**20b** Green-brown, most easily seen in the larger filamentous forms, which have no cross walls along the filaments. Can easily be conused with green algae (chlorophytes) especially older cells, but does not stain black with iodine solution **Xanthophytes** (Harosa, Xanthophyta)

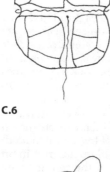

**C.6**

**21a** Single cell, basically brown but may be tinges of red, blue or green with a tinge of brown, with a distinct groove encircling the cells (C.6), in which there is a flagellum and a second flagellum trailing downwards, but flagella may be very difficult to see. Distinct wall plates may be visible covering the cell **Dinoflagellates** (Pyrrophyta, Alveolata)

**21b** No distinct groove around cell **22**

**22a** Single cell, colonies or filaments with individual cells circular, needle-like, moon-shaped, s-shaped or boat-shaped with distinctive patternings in the walls, which are of silica, best seen in dead cells **Diatoms** (Bacillariophyta, Harosa)

**22b** Not like this **23**

**23a** Flagellated cell **24**

**23b** Not flagellated **25**

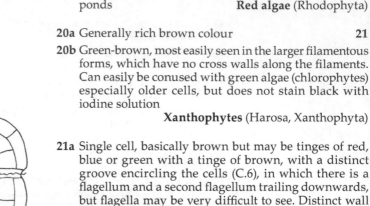

**24a** Small flagellate, covered in organic or siliceous scales (which may be visible as a roughness on the surface (C.7)); usually single celled but may be colonial **Chrysophytes** (yellow-green algae, yellow algae, golden algae, Chrysophyta, Harosa)

**C.7**

**24b** Small flagellate with two flagella, covered in scales and with an additional projection, the haptonema (C.8) between the flagella **Haptophytes** (Haptophyta, Harosa)

**25a** Macroscopic filaments or more solid plants. Scarce in freshwaters, but common in brackish waters and the sea **Brown algae** (Phaeophyta, Harosa)

**25b** Appearing as rust coloured brown streaks where water seeps from rocks or sometimes clay sediments and banks. Colourless fine filaments but surrounded by sheaths and deposits of brown iron oxide **Iron bacteria**

**C.8**

Key C will allow you to go to the more detailed published keys for protozoans and algae, but sometimes these are very expensive. I have thus provided three keys in this chapter: to the macroscopic algae and algae that are prominent en masse, though a microscope will be needed for their identification (Key D); to microscopic algae (including cyanobacteria) that occur on sediments and attached to surfaces such as plants or rocks (Key E); and to the colourless protozoans (Key F), which are also microscopic. D and E are keys to the more widespread genera. F goes to orders and sub-orders. Living, not preserved, material is required and I have included figures that illustrate possibly unfamiliar terms, though I have tried to use familiar rather than formal terms. Once you have arrived at a name using the keys, enter it into an internet search engine and you will usually be able to access dozens of images to check your identification. But be careful. There are many errors on the internet. Always go back to the original page of the image and check out several images for consistency. Learned societies concerned specifically with the groups (for example the British Phycological Society's AlgaeBase (www.algaebase.org)) are generally very reliable, but the web sites of individuals may not be. There are about 3,000 taxa of algae in Britain and Ireland in about 1,400 genera, about 200 of which may be planktonic and 1,200 benthic (occurring on the mud or on surfaces). Protozoa are almost as diverse and it is possible that you will find genera not in the key. However, the keys will bring you into the region of similar organisms based on colour and overall structure. You should then be able to find the correct genus and species by consulting the standard works (see Chapter 10). But sometimes there is uncertainty even among expert taxonomists, especially for the cyanobacteria! At the stage of determining species, measurements of size may become crucial and you will need an eyepiece micrometer calibrated against a reference scale. In the Keys here I have not indicated precise sizes because different species within a genus may vary greatly in size.

# Key D Algae visible, at least *en masse*, to the naked eye

**1a** Without flowers or conventional flat leaf blades, but distinctly plant-like (Fig. 3.2) with an upright stem, usually up to 20 cm (but can be much larger) and branches. Dark green or greyish, sometimes evergreen. There may be a particular garlicky smell. The branches are whorled around nodes and the stem comprises a sequence of very large single cells. Some species have a rough feel to the touch, but others not. Sometimes bright orange reproductive organs are visible among the distal branches. Identification, even to genus, needs fine characters, a specialist terminology and a detailed key. There are five genera in all

**Stoneworts, Charophyta, 2**

**1b** Lettuce-like, tubular, or in small tufts anchored to rocks or the bottom                                                                **3**

**1c** In gelatinous masses, anchored or rolling free          **6**

**1d** Crusts on rocks or other stable surfaces                     **7**

**1e** Green flocs, like masses of cotton wool in the water  **9**

**1f** Small (up to 7 mm ) rounded dark-green lumps with rhizoids anchoring them into sediment, like tiny balloons (D.1)                          ***Botrydium*** (xanthophyte)

**D.1**

**2a** Stem cells covered by additional long thin cells (corticated), extending usually in parallel over the surface (D.2 left); branches usually themselves branched with whorls in a similar manner to the main stem. Greyish, rough feel and distinctive smell
***Chara*** (green alga)

**2b** Green, not corticated (D.2 right), smooth to the touch, branchlets divided in forks          ***Nitella*** (green alga)

**2c** Green, not corticated (D.2, right), smooth to the touch, branchlets undivided          ***Nitellopsis*** (green alga)

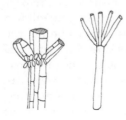

**D.2**

**mucilaginous**
producing mucilage
(a viscous, sticky
substance)

**3a** Flat and lettuce-like or tube- or intestine-like, with flabby consistency                                                               **4**

**3b** Tufts on sediment or hard surfaces                             **5**

**4a** Flat sheets, green, usually up to 20 cm in extent
***Monostroma*** (green alga)

**4b** Intestinal in form, often 20 cm or more long, and in considerable masses supported in the water by contained gas    ***Ulva*** (often called ***Enteromorpha***) (green alga)

**5a** Gelatinous tufts, brown or green, mucilaginous, and with complex structure (cells in several parallel layers, connected by threads) under the microscope. Whorls of branches give a beaded appearance to the eye. More

common in streams (as with all the freshwater red algae) but may occur in rocky tarns
*Batrachospermum* (red alga)

5b Feathery tufts, deep brown, without a beaded appearance, distinctive foetid smell, especially when crushed between the fingers *Hydrurus* (yellow-green alga)

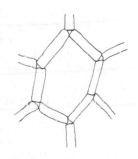

6a Gelatinous clumps, grass green in colour, with two long hairs emerging from each cell when examined under the microscope *Tetraspora* (green alga)

6b Brownish or dark greenish blobs, often anchored to the bottom, cells very small, spherical or slightly ovate, all of one kind in masses of mucilage *Aphanocapsa* (cyanobacterium)

6c Spherical or irregular lumps lying free but may be anchored. Under the microscope (D.3), the blobs are made up of filaments with two kinds of cells. The more abundant are pigmented, the less abundant are larger and paler (heterocysts). Nitrogen fixation occurs in these *Nostoc* (cyanobacterium)

**D.3**

7a Red or magenta crust, squarish cells with one red chloroplast per cell *Hildenbrandia* (red alga)

7b Brown crust 8

8a Crust with distinct margins of short brown filaments, anchored by basal rhizoids that are lightly coloured. Several brown chloroplasts per cell *Heribaudiella* (brown alga)

8b Crusts of several layers, but without colourless rhizoids. One chloroplast per cell. Most common in streams *Phaeodermatium* (yellow-green alga)

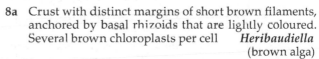

**D.4**

9a Cells arranged in green nets around polygonal meshes (D.4), with sides and meshes often visible to the naked eye *Hydrodictyon* (green alga)

9b Filaments, generally straight 10

10a Filaments unbranched or rarely branched 11

10b Filaments profusely branched 17

11a Filaments en masse feel slimy, and under the microscope have mucilage layers covering the walls (detectable by adding a very small drop of Indian ink) 12

11b Filaments not slimy 13

**12a** Green, chloroplasts usually two per cell, star-shaped
*Zygnema* (green alga)

**12b** Green, chloroplasts in helices (like the structure of DNA) *Spirogyra* (green alga)

**12c** Green, chloroplasts flat or slight twisted plates
*Mougeotia* (green alga)

**13a** Dull green, not staining black with iodine solution **14**

**13b** Bright grass green (though paler in old and decrepit filaments when cell will stain black with iodine solution) **15**

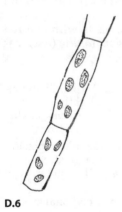

D.5

**14a** Filaments without cross walls; occasional branching; sexual organs (large rounded egg-bearing structures, and short side branches (D.5) close to them bearing male cells) not uncommon *Vaucheria* (xanthophyte)

**14b** Walls constructed in two halves, thickest at the ends, close to the cross walls, thinnest in the middle of the cell where the pieces overlap (D.6). At the ends of the filament, an empty half-wall is often seen *Tribonema* (xanthophyte)

**15a** Chloroplast wrapped around the inside wall of the cell, like a piece of flattened chewing gum. This arrangement is described as parietal (D.7) **16**

**15b** Several ribbon-like chloroplasts running the length of the cell *Sirogonium* (green alga)

**15c** Coarse green filaments, sometimes sparsely branched, with many small chloroplasts. Rough to the feel and often profusely covered with periphyton
*Cladophora* (green alga)

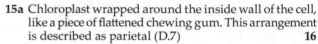

D.6

**16a** Attached by a basal cell to a surface, though often the filaments break free and float in the water buoyed by gas bubbles *Ulothrix* (green alga)

**16b** Free-floating. (In practice may be indistinguishable from *Ulothrix*) *Klebsormidium* (green alga)

**17a** Filaments without colourless spines or hairs **18**

**17b** Spines or projections on the cells or filaments ending in colourless hairs **19**

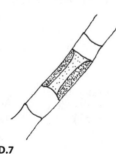

D.7

**18a** Filaments occurring in dense felts, and sometimes rolled into free-living balls by wave action
*Aegagropila* (green alga)

**18b** Coarse green filaments, sometimes sparsely branched, with many small chloroplasts. Rough to the feel and often profusely covered with other small algae and bacteria (periphyton) *Cladophora* (green alga)

**18c** Filaments permeating the surfaces of wood, limestone, or shell, irregularly branched. May be visible en masse, but microscopy often needed for detection
*Gomontia* (green alga)

**D.8**

**19a** Filaments or branches ending in a colourless hair, but without spines on the cells **20**

**19b** At least some cells bearing spines or colourless projections **21**

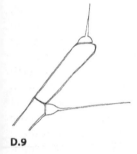

**D.9**

**20a** Cells have a series of lines or scars at one end (D.8), resulting from previous cell divisions. Egg bearing spherical structures are often present
*Oedogonium* (green alga)

**20b** Green feathery structure. Prominent primary filaments have tufts of branches ending in an erect hair. Filaments are attached to the bottom by colourless rhizoids
*Draparnaldia* (green alga)

**20c** Green, mucilaginous with prostrate colourless filaments attached to a surface, and erect green filaments that are branched or whorled, but main axis is not prominent
*Stigeoclonium* (green alga)

**D.10**

**21a** Most cells have a colourless spine (seta) with a bulbous base (D.9) *Bulbochaete* (green alga)

**21b** Branched erect filaments emerge from a flat green plate of cells adherent to a surface and from which a few colourless spines emerge (D.10). Sometimes the erect portion breaks away leaving just the plate
*Coleochaete* (green alga)

## 3.7 Attached and free-living periphyton

Periphyton means 'around the plants' but its meaning has been corrupted to include the communities of microorganisms that are attached to plants, live loosely around them, are attached to animals, rocks, wood and any other underwater surface, and move over and in sediments. There are specific terms for each sub-habitat (epiphytes, epizoon, epilithon, epixylon, epipsammon, epipelon (for associations with plants, animals, rocks, wood, sand grains and mud, respectively, but the scientific world is generally lazy about using them and periphyton does for all. Sometimes the term benthic is used to distinguish all other communities from plankton, but it really means 'bottom-living'. There are many algae that are found in more than one habitat, for example sediment-livers that become displaced into the plankton, or which entangle themselves among the epiphyton. Plankters may be found on the plants, rocks or sediments when they sink out of the water. It is thus worth using all of Keys E and F, below, and H (in Chapter 4) to come to a preliminary identification and to use the internet to see images to help check. You may then have to go to more detailed works to be certain, but for many purposes these keys should allow you to gain a substantial ability at identification and confidence to use the more detailed literature.

Key E is to the periphyton (in its broad sense). Many of the algae in this key are diatoms, for which identification is much easier if the wall markings are made clear. To do this, part of the sample needs to be cleaned by warming and oxidising in hydrogen peroxide. The latter is available domestically. Use as strong a solution as available (usually 20%) and after very gentle heating on a hot plate or kitchen cooker, leave to stand for several hours with more hydrogen peroxide added. Wear safety spectacles and latex gloves in case you overheat it and the liquid boils. Hydrogen peroxide can cause painful burns if it gets onto the skin. Then settle or centrifuge the diatoms, pour off the supernatant water and resuspend in a small volume of distilled water. Place a microscope cover glass on a glass plate (a microscope slide will do), made a little greasy by handling, and pipette on a volume of the suspension until it bulges with surface tension at the edges. If the glass is too clean, the drop may overrun the edges; if slightly greasy, it will not. Allow to dry, which will take several hours or overnight, then place a small drop of high refractive index mountant (Naphthrax or similar; see list of equipment suppliers in Chapter 10) on a clean dry microscope slide and use the sticky drop

**plankter**
an individual organism in the plankton

**supernatant**
the liquid above a solid residue

**Fig. 3.9** When cleaned and mounted on a slide, diatoms appear much as on the left. Some cells fall apart so that there is a lot of silica debris among cells that are recognisable. The sample was from a sediment core and many cells break apart because of abrasion and passage through the guts of worms. The two halves of the sarcophagus-shaped cell (*Gomphonema* species) at the centre right are coming apart because the girdle bands that bind them together in life have fallen away. The hairpin-shaped object running to the bottom right corner is a girdle band from a larger species. The right hand photograph (by Rakel Gudmundsdottir) was taken with a scanning electron microscope and shows a cell (*Navicula* species) that has remained intact. Fragments of a broken cell lie to its left, but one of the valves (with markings (striae)) is clearly shown with a raphe running along the centre. A girdle band runs around the sides. Underneath, and not visible, will be the second valve, which in this case will closely resemble the one that is visible.

to pick up the cover glass. Warm very gently on a hotplate or electric stove until the solvent, in which the mountant is dissolved, evaporates. There will be a frothy bubbling and when this ceases and becomes a slow bubbling the process is complete and you should not heat any more for fear of burning the mountant and cracking the slide. When the slide is cool, examine under high power or, if available, oil immersion lenses. The markings should be clear. You will need to match up the living and cleaned cells by comparison of shapes, but remember that the orientation in living material may differ from what you see on the slide, where the walls break apart into the valves (the flat side) and the girdle bands which bind them together like tape on a parcel (Fig. 3.9). Some valves may appear in side view, which is generally rectangular and this is sometimes the most common view in live material. An excellent key to common river diatoms, which include many pond forms, and more hints are available in Kelly (2000).

## Key E Periphyton, both attached to surfaces and free living

**1a** Firmly attached to surfaces by a stalk, pad or basal cell, in a mass of mucilage, or flush with the surface   **2**

**1b** Free-living   **17**

**2a** Attached by a stalk   **3**

**2b** Attached by a pad, basal cell, flush with the surface, or in a mass of mucilage   **4**

**3a** Stalks branched; green cells held at the ends of gelatinous stalks (E.1), attached to filamentous algae or animals   *Colacium* (euglenoid)

**3b** Stalk unbranched, cells brown and wider towards one end, then usually tapering and always tapering towards the narrower end. In cleaned material (see p. 52) the markings (striae) run across the cell to a central line (the raphe) in front (valve) view, and there are one or two prominent dots (stigma) to one side of the middle (E.2, left). In side (girdle) view the cell also tapers and the girdle bands run centrally along the cell, joining the two valves (E.2, right). This genus can also be motile and free-living   *Gomphonema* (diatom)

**4a** Cells attached flat to the surface, brown, oval in shape. In cleaned material there is a raphe on one side but not the other and the markings radiate out to the edges   *Cocconeis* (diatom)

**4b** Cells attached at one end, sometimes on a basal pad  **5**

**5a** Cells small and contained in a vase or lorica directly attached to a surface   **6**

**5b** Cells not so contained   **7**

**5c** Cells in a mucilaginous colony of 2–8 cells, with ovoid or sausage-shaped blue-green bodies inside (these are called cyanelles and are symbiotic cyanobacteria which function as chloroplasts). Two long erect gelatinous hairs emerge from each cell   *Gloeochaete* (glaucophyte)

**6a** Cells in an open vase, may be single or sometimes in small colonies. A small brown flagellated cell is contained to one side of the vase (E.3). The structure is very like that of the free-living *Dinobryon* (see Key H)   *Epipyxis* (yellow-green alga)

**6b** Spherical lorica, containing brownish cells, with extensions burrowing into the host surface and some long thin projections emerging from a pore at the top (E.4)   *Chrysopixis* (yellow-green alga)

E.1

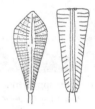
E.2

**raphe**
a mucilage-producing slit in the valves (the two halves of the silica cell wall) of many diatoms

E.3

E.4

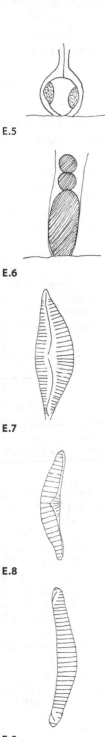

E.5

E.6

E.7

E.8

E.9

**6c** Bottle-shaped lorica, with a narrow neck (E.5), inside which is a brown cell *Lagynion* (yellow-green alga)

**7a** Cells blue-green, dull green or blackish green, not staining with iodine **8**

**7b** Green, filamentous, branched or unbranched, attached at one end or from a basal pad. Older cells (but not young, fresh green ones) will stain black with iodine solution **see Key D to macroscopic algae**
All of the green filamentous genera there may occur on surfaces in small numbers so that they are not macroscopic en masse

**7c** Grass green, older cells staining black with iodine. Single cells, oblong, spindle-shaped or ovate, sometimes with a point at the upper end. Attached by a very short stalk or pad. One or perhaps two chloroplasts *Characium* (green alga)

**7d** Cells brown **9**

**8a** Single, elongate cell, budding off rounded spores at the top (E.6). There may be a sheath surrounding the cell and spores *Chamaesiphon* (cyanobacterium)

**8b** Sausage-shaped cells, sometimes with a short spine or point at the distal end *Ophiocytium* (xanthophyte)

**8c** Cells oblong, spindle-shaped or ovate, sometimes with a point at the upper end. Attached by a very short stalk or pad. More than one chloroplast and does not stain black with iodine solution. Can easily be confused with the very similar green alga, *Characium*, which does stain *Characiopsis* (xanthophyte)

**9a** Cells symmetrical about the mid point and curved like a banana, or less-than-full moon or with one edge flat and the other curved **10**

**9b** Cells some other shape **12**

**10a** Several cells contained in a mucilage tube attached to a surface *Cymbella* (diatom)

**10b** Cells not in a mucilage tube **11**

**11a** Cells with a longitudinal raphe positioned towards the inner edge of the curved cell. The raphe may have a shallow S-shape (E.7) *Cymbella*
(and *Encyonema*, which is only subtly different (see Kelly, 2000) (diatoms)

**11b** Cells with coarse markings (visible in cleaned material) and a prominent V-shaped raphe (E.8) *Epithemia* (diatom)

**11c** Cells with very short raphes, difficult to see, at either end (E.9). Mostly in acid waters *Eunotia* (diatom)

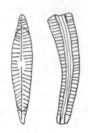

**E.10**

**E.11**

**E.12**

**E.13**

**E.14**

**12a** Wedge shaped in both side (girdle) and valve view, with the wide end square in girdle view, but rounded in valve view (E. 10)  **13**

**12b** Not wedge shaped in either view  **16**

**13a** Cell prominently bent in girdle view
*Rhoicosphenia* (diatom)

**13b** Cell not bent in girdle view and manikin-shaped, with a head and tapering body (E.11), in valve view  **14**

**13c** Cells simply wedge-shaped with rounded upper ends in valve view, no raphe and coarse costae extending across the valve view (E.12). Very small, often attached to sand grains  ***Martyana***
(formerly *Opephora*) (diatom)

**14a** With a raphe extending along the central axis of the cell (E.13)  **15**

**14b** Without a raphe, cells in fan-shaped, or spiralling colonies, like pencil sharpenings. The head of the manikin is present but not very prominent
***Meridion*** (diatom)

**15a** Large cells, usually attached to rocks, and associated with much mucilage ('rock snot'). Several coarse stigmata ('beauty patches') (E.14) to one side at the middle of the cell in valve view
***Didymosphaenia*** (diatom)

**15b** Cells not especially coarse, with up to two stigmata at centre of cell. This genus may also be attached by a stalk or free-living (see E.2)  ***Gomphonema*** (diatom)

**16a** With a raphe on one valve, but not on the other
***Achnanthes*** (diatom)
(this genus has been extensively divided into several new ones on the basis of electron microscope observations; the new genera (for example, *Karayevia, Kolbesia, Psamothidium, Achnanthidium, Planothidium, Reimeria* and *Rossithidium* are recognised in the key by Kelly (2000))

**16b** Cells attached to the surface at one corner. There may be a zig-zag line of several of them, attached to one another at the corner. Septa extend part way across the cells (E.15) and are most visible in cleaned material. This genus can also be planktonic, forming star-shaped colonies with the cells attached to one another at the corners  ***Tabellaria*** (diatom)

**16c** Cells generally much longer than broad, sometimes needle-like, sometimes spindle-like, sticking out from a surface like spines, but always singly. No raphe (E.16), but cleaned material and oil immersion lenses may be needed to confirm this  ***Synedra*** (diatom)

**17a** Free-living, unicellular or colonial, moving with flagella **Check Key H** (many flagellates can be both planktonic and loosely associated with surfaces)

**17b** Free-living, non-motile or moving slowly by gliding **18**

**18a** Cells dull green, not staining with iodine solution, with many projections but not prominently divided across the middle or triangular in side view *Pseudostaurastrum* (xanthophyte)

**18b** Blue-green, purple, pink, reddish or deep brown and then forming a colony **19**

**18c** Cells grass green, or staining black with iodine solution **34**

**18d** Cells bright brown or yellowish (but not staining black with iodine solution) and contained within a cell wall that has geometric markings on it (diatoms) **55**

**19a** Single cells (or double if dividing) **20**

**19b** Colonial, with filaments or individual cells enclosed in mucilage **21**

**19c** Filamentous but not enclosed in a mass of mucilage **26**

E.15

**20a** Cells large, up to 50 μm, surrounded by a prominent sheath (E.17) often showing striations *Chroococcus* (cyanobacterium)

**20b** Cells small (<5 μm) spherical or elongate, without sheaths, *Synechococcus* (cyanobacterium)

**21a** Colony of filaments surrounded by mucilage **22**

**21b** Colony of individual rounded cells **23**

**22a** Filaments not branched, within a mass of mucilage, the whole forming a rounded body like a pea or plum, or a flattened sheet, brownish green in colour. Filaments have two sorts of cells, the least common (heterocysts) being larger and paler (see D.3). Nitrogen fixation goes on in these cells *Nostoc* (cyanobacterium)

E.16

**22b** Filaments radiating out from the base, within mucilage, with a heterocyst at one end and tapering to a colourless hair at the other. The filaments do not branch but they are contained within sheaths that may branch *Rivularia* and *Calothrix* (when sheaths are not branched) (cyanobacteria)

**23a** Cells distributed irregularly in mucilage **24**

**23b** Cells in a geometric squarish or cubical pattern **25**

E.17

**24a** Colony with at least 16 small, rounded cells, usually many more, each pigmented throughout (no distinct chloroplast)         *Aphanocapsa* (cyanobacterium)

**24b** Colonies of a few cells, in a mass of mucilage with individual sheaths around each cell. Sometimes red in colour         *Gloeocapsa* (cyanobacterium)

**24c** Colony of 4–16 spherical or ellipsoid cells, each with several sausage-shaped chloroplasts twisting around one another. The 'chloroplasts' are symbiotic cyanobacteria         **Glaucocystis** (glaucophyte)

**25a** Colony a flat sheet, with cells arranged in rows and columns         *Merismopedia* (cyanobacterium)

**25b** Colony a three-dimensional cube with cells arranged in rows, columns and ranks         *Eucapsis* (cyanobacterium)

**26a** Individual filaments not enclosed in sheaths (check also Key H)         **27**

**26b** Individual filaments or groups of filaments enclosed in a sheath         **30**

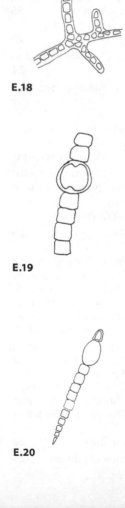

**E.18**

**27a** Filaments branched, rather irregularly, of various lengths and emerging at various angles (E.18). Furry growths on wet rocks, especially where prone to drying out         *Stigonema* (cyanobacterium)

**27b** Filaments not branched         **28**

**28a** Two or three sorts of cells present         **29**

**28b** All cells basically similar, with some modification of shape in the terminal cells. Filaments straight         *Oscillatoria* (cyanobacterium)

**28c** All cells similar. Filaments helical. Cell cross walls obscure         *Spirulina* (cyanobacterium)

**E.19**

**28d** All cells similar. Filaments helical. Cell cross walls distinct         *Arthrospira* (cyanobacterium)

**29a** Two sorts of cells present with larger, paler heterocysts spaced along the filament or located at one end (E.19)         *Anabaena*

**29b** Filaments have a heterocyst at one end, backed by a much larger cell, the akinete (which is a spore) and then a chain of vegetative cells gradually tapering to a colourless hair (E.20)         *Cylindrospermum* (cyanobacterium)

**30a** Sheaths around individual filaments         **31**

**E.20**

**30b** Several filaments in a bundle surrounded by a sheath         *Microcoleus* (cyanobacterium)

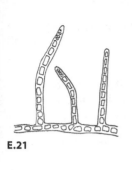

**E.21**

**31a** Filaments unbranched without heterocysts **32**

**31b** Filaments branched or unbranched, with heterocysts **33**

**32a** Filaments tending to occur singly with a sheath often protruding beyond the filament
*Lyngbya* (cyanobacterium)

**32b** Filaments tending to occur in gelatinous or leathery masses, sometimes floating up, buoyed by gas bubbles, from the bottom to the surface
*Phormidium* (cyanobacterium)

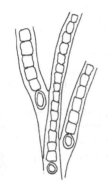

**E.22**

**33a** Filaments branched, with branches emerging at right angles (E.21) to main axis
*Hapalosiphon (Fischerella)* (cyanobacterium)

**33b** Filaments unbranched and terminating in a heterocyst, but sheaths branched (E.22)
*Tolypothrix* (cyanobacterium)

**34a** Green algae, with cells not prominently divided in the middle by an isthmus or prominent line **35**

**34b** Cell with a prominent isthmus or line across the middle that divides them into two halves (E.23) with a single chloroplast in each half. Walls often ornamented (placoderm desmids). Other than a very few species of *Cosmarium, Staurastrum* and *Closterium*, desmids are found in acidic, low conductivity waters **36**

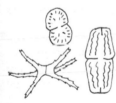

**E.23**

**35a** Cell not divided into two symmetrical halves
**check Key H**
(for *Scenedesmus* and related non-motile green genera)

**35b** Cell divided into two symmetrical halves, with a chloroplast at each end, but without a prominent, deep isthmus or line and with usually plain, unornamented walls (E.24) (saccoderm desmids). **50**

**36a** Cell single **37**

**36b** Cells in filaments **46**

**36c** Cells cottage-loaf shaped and held in a colony with mucilage     *Cosmocladium* (desmid, green alga)

**37a** Cell long and thin, more than four times as long as broad **38**

**37b** Cell no more than three times as long as broad **42**

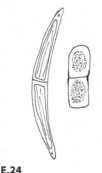

**E.24**

**38a** Cell curved in an arc, often prominently
*Closterium* (desmid, green alga)

**38b** Cell straight **39**

**E.25**

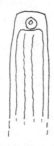

**E.26**

**39a** Cell suddenly widens (like a life-saving ring) (E.25) to either side of the isthmus **40**

**39b** No such widening **41**

**40a** Widening without teeth, chloroplast not ridged
*Pleurotaenium* (desmid, green alga)

**40b** Widening has a line of teeth along its ridge, making it look like a cog wheel
*Docidium* (desmid, green alga)

**41a** Ends of the cells have a notch, wall does not have ornamentations *Tetmemorus* (desmid, green alga)

**41b** Walls with pores, chloroplast ridged and a vacuole with a crystal (E.26) that jiggles at either end
*Penium* (desmid, green alga)

**41c** Walls with pores, chloroplast ridged, but no crystals
*Haplotaenium* (desmid, green alga)

**42a** Cell without spines, blunt projections or deep indentations, other than the central isthmus, cottage-loaf shaped *Cosmarium* (desmid, green alga)

**42b** Cell with spines or blunt projections or deep indentations **43**

**43a** Cell with spines **44**

**43b** Cell with blunt projections or deep indentations **45**

**E.27**

**44a** Generally four spines, one from each corner, cell flattened *Staurodesmus* (desmid, green alga)

**44b** Many spines, usually at least 10 per half cell; cell not flattened *Xanthidium* (desmid, green alga)

**45a** Cell with a deep indentation at the poles, and blunt projections (E.27). Slightly flattened
*Euastrum* (desmid, green alga)

**45b** Cell greatly flattened, symmetrical with many indentations at the edges giving the shape of a complex star
*Micrasterias* (desmid, green alga)

**45c** Cell not flattened but with a triangular cross section and projections, often long, and often with short subprojections *Staurastrum* (desmid, green alga)

**46a** Filament not twisted **47**

**46b** Filament longitudinally twisted with a line of projections running helically along it (E.28)
*Desmidium* (desmid, green alga)

**E.28**

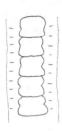

**E.29**

**47a** Filament surrounded by a thick layer of mucilage (highlight with a drop of Indian ink); cells only shallowly indented and lacking projections (E.29)
*Hyalotheca* (desmid, green alga)

**47b** Filament may be mucilaginous but the layer is much thinner than the thickness of the cells **48**

**48a** Cells barrel-shaped with only a small isthmus
*Bambusina* (desmid, green alga)

**48b** Cells cottage-loaf shaped **49**

**49a** Cells smooth (not ornamented) but with granules within the cells at either end
*Teilingia* (desmid, green alga)

**49b** Cells smooth but with small peg-like projections where they adjoin one another in the filament
*Sphaerozosma* (desmid, green alga)

**49c** Cells smooth and flattened in edge view
*Spondylosium* (desmid, green alga)

**50a** Cell relatively long and thin, more than five times as long as broad **51**

**50b** Cell relatively short and fat, less than five times as long as broad **52**

**51a** Cell slightly curved, with flattened ends
*Roya* (desmid, green alga)

**51b** Cells prominently curved with pointed ends
*Closterium* (desmid, green alga)

**51c** Cell not regularly curved, very long and thin with wall markedly granular or with short spines
*Gonatozygon*
(*Genicularia* is less thin, more granular with spiralling chloroplasts) (desmid, green alga)

**52a** Chloroplasts star-shaped **53**

**52b** Chloroplasts spiral or flat plates **54**

**53a** Cell with slight narrowing at middle, fine pores in wall *Actinotaenium* (desmid, green alga)

**53b** Cell not narrowed at middle
*Cylindrocystis* (desmid, green alga)

**54a** Chloroplast a flat plate, ends of cell rounded
*Mesotaenium* (desmid, green alga)

**54b** Chloroplasts cylindrical with prominent longitudinal ridges. Crystals contained in vacuoles at the ends of the cells *Netrium* (desmid, green alga)

**54c** Chloroplast spiral *Spirotaenia* (desmid, green alga)

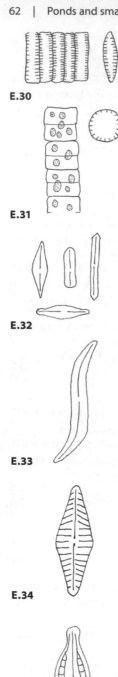

**E.30**

**E.31**

**E.32**

**E.33**

**E.34**

**E.35**

**55a** Single cells (the remainder of the key includes diatoms and you will need cleaned (see p. 52) as well as live material to be sure of your identification **57**

**55b** Filaments **56**

**56a** Cells long and thin and joined into a filament like the palings in a fence (E.30) *Fragilaria* (this diatom genus has been greatly revised on the basis of electron microscopic features. The main new genera (*Pseudostaurosira*, *Staurostirella*, and *Fragilarioforma*) are included in the key by Kelly (2000)

**56b** Cells short and fat, circular in cross section and resembling short, fat, gelatin pill capsules with a line down the middle where the two valves (frustules) abut (E.31) *Melosira* (diatom)

**57a** Cell boat-shaped (E.32) (with a little imagination) and with bilateral symmetry both end to end and side to side. A single, simple raphe runs centrally along the full length of the valve, with an interruption at the middle. The side (girdle) view is rectangular and not much help **58**

**57b** Cell S-shaped (E.33), usually large and with criss-cross striae forming a square mesh. Raphe not in a ladder like structure *Gyrosigma* (diatom)

**57c** Cell with no raphe (or at least not one that is obvious), two raphes, a raphe in a ladder-like structure or a very short one at either end. Cells may be boat shaped or not, but there is not a single central raphe **64**

**58a** Cell with striae to either side of the raphe that are fine, and radiate, sometimes turning towards the poles at the distal ends. Central area without striae not large. Striae extend to close to raphe. Central nodules of raphe are not turned to the side (E.34)

*Navicula*
(this genus has been extensively revised, following electron microscope examination, into several new genera: *Aneumastus*, *Cavinula*, *Craticula*, *Ctenophora*, *Diadesmis*, *Fallacia*, *Luticola*, *Sellaphora*, *Cosmioneis*, *Petroneis*, *Placoneis*). These are included in Kelly's (2000) key

**58b** Cell has more complex structure, with greater patterning in the wall markings, or turned ends in the raphe in the central area **59**

**59a** Prominent cross-shaped area, lacking striae, in the centre *Stauroneis* (diatom)

**59b** Prominent cross-shaped area absent **60**

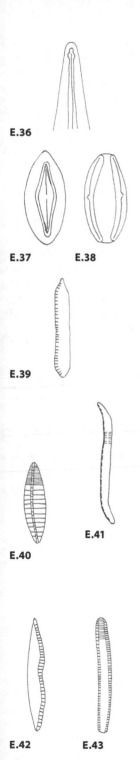

E.36

E.37    E.38

E.39

E.40    E.41

E.42    E.43

**60a** Striae obviously formed of series of dots rather than lines                                    *Anomoeoneis* (diatom)

**60b** Striae of thick ribs, large cells, central raphe nodules turned the same way, generally wide margin between the area covered by striae and the raphe
*Pinnularia* (diatom)

**60c** Striae very difficult to see in detail, but edge of valve has a broad clear area        *Brachysira* (diatom)

**60d** Striae clearly visible and of thin lines                          **61**

**61a** Central nodules of raphe turned in the same direction                                  *Caloneis* (diatom)

**61b** Central nodules of raphe turned in opposite directions                                *Neidium* (diatom)

**61c** Central nodules of raphe not turned              **62**

**62a** Chambers like blisters (E.35), along each side of the valve                                  *Mastigloia* (diatom)

**62b** Chambers absent                                        **63**

**63a** Raphe paralleled by a thickened, plain area (E.36), but otherwise appearance of striae is uniform. Always in acid habitats                            *Frustulia* (diatom)

**63b** Striae change pattern close to raphe; there is a plain structure paralleling the raphe and a large central bare area (E.37). Cells generally ovate but can be constricted in the middle like a violin        *Diploneis* (diatom)

**64a** Two longitudinal raphes present each to one side of the midline (E.38)                        *Amphora* (diatom)

**64b** Raphe extending the length of the valve in a ladder-like structure (E.39), at the edge or offset from the middle
**65**

**64c** Raphe short at either or one end of the cell      **66**

**64d** No raphe apparent                                      **67**

**65a** Raphe to one side of the midline but not at the edge of the valve. Thick costae extend across the valve (E.40), in addition to thin striae          *Denticula* (diatom)

**65b** Valve sigmoid, raphe at the edge (E.41)
*Stenopterobia* (diatom)

**65c** Raphe at the same edge of the cell in both valves, cells often slightly curved (E.42)        *Hantzschia* (diatom)

**65d** Raphes on different edges in the two valves (E.43), edge usually straight                          *Nitzschia* (diatom)

**E.44**

**E.45**

**E.46**

**E.47**

**66a** Boat-shaped, with pointed ends, raphe extending a short way from either end but far from the middle, and surrounded by a clear area (E.44). Striae extremely fine and usually not visible    *Amphipleura* (diatom)

**66b** Cell symmetrical in the longitudinal plane, but not in the transverse, with a manikin shape having one tapering end and a rounded 'head' at the other. Short raphes, turning to one side, extend from each end and are not very long on one valve (E.45), short and barely visible on the other    *Peronia* (diatom)

**67a** Cell large, nearly circular but indented in the middle so that it forms the shape of a saddle
*Campylodiscus* (diatom)

**67b** Cell twisted like a skein of wool or rope
*Amphiprora* (diatom)

**67c** Cell curved with central swollen plain bulge to the inner side    *Hannaea* (diatom)

**67d** Cell shaped like a regular jigsaw puzzle piece with rounded ends and a central squarish area. Septa and costae extend across the valve (E.46)
*Tetracyclus* (diatom)

**67e** Cell large, slipper-shaped (wider close to the pointed ends, narrower at the middle) with wide striae at the margins and an undulating surface like a switchback ride, so that the whole cannot be in focus at the same time    *Cymatopleura* (diatom)

**67f** Cell large and looks as if it is falling apart with the two valves, each the shape of a strung bow, separated by many bands (E.47)    *Rhopalodia* (diatom)

**67g** Cell large, pointed or ovate with large marginal striae and the surface in a flat plane    *Surirella* (diatom)

## 3.8 Protozoa

The old categories of algae (photosynthetic) and protozoa (not photosynthetic) were discontinued long ago, insofar as an understanding of relationships and evolution was concerned. There are closely related genera that differ only in the presence or absence of chloroplasts and modern work on genomes has demonstrated unexpected links and differences. But the old, artificial distinction is still useful in the business of identification. Key F is to the colourless (though in fact many shades of grey) eukaryotes once classed as Protozoa. It is not comprehensive for genera and is a key to orders and sub-orders for the most part, with the more common genera indicated. Its role is to give an introduction that might stimulate further interest. Many people ignore the protozoans in favour of the more colourful algae and there is only a handful of expert protozoologists left. But the protozoans are nonetheless very important in the functioning of ponds. The key should be used in tandem with the internet to arrive at a tentative identification. It is simplified from Jahn *et al.* (1978), which also includes many orders of parasitic protozoa that are not included here. There is no single authoritative key to protozoan genera or species, but if you wish to specialise in this group, Patterson (2003) is also very useful, if expensive. One problem with protozoans is that they do not preserve easily, and many tend to contract or burst when chemicals such as iodine solution are added. It is a good idea to make drawings from living material to help in identification. Indeed this is always a good idea.

**cilium (plural cilia)**
hair-like structure that projects from the cell surface and can move

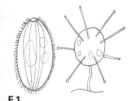

**F.1**

**pseudopodium (plural pseudopodia)**
an extension of the cell surface as the cell moves

**flagellum (plural flagella)**
long whip-like appendage used for swimming

## Key F Major orders of free-living protozoa

**1a**  Cells with cilia or sucking tentacles (F.1)

           **Ciliophora, 3**

**1b**  Without cilia or sucking tentacles    **2**

**2a**  With one or more flagella and with or without pseudopodia and amoeboid movement

           **Mastigophora, 18**

**2b**  With pseudopodia and amoeboid movement but without flagella

      **Sarcodina** (*Rhizopoda, Amoebozoa*), **25**

**3a**  Cilia always present        **Ciliata, 4**

**3b**  Cilia present only in early stage but adult has tentacles sticking out from the cells like bunches of spines with a slight swelling at the tip (a group found attached to filaments or on the surfaces of animals like turtles)   **Suctorea**
    (*Tokophyra, Podophyra, Acineta, Anarma, Squalophyra*)

**F.2**

**F.3**

**F.4**

**F.5**

**F.6**

**4a** Cilia of the same type over the surface; no zone of thicker fused cilia around the mouth
**Holotrichida, 7**

**4b** Thicker, fused cilia forming a zone around the mouth, along which waves of movement pass **5**

**5a** Zone of fused cilia winds counter-clockwise (F.2) around mouth, cells mostly on stalks **Peritrichida, 17**

**5b** Zone of fused cilia winds clockwise (F.3) around mouth **6**

**6a** Mouth at the base of a deep bell-shaped funnel, often spiralled; attached to other animals
**Chonotrichida** (*Spirochona*)

**6b** Mouth not at the base of a bell-shaped funnel
**Spirotrichida, 13**

**7a** Mouth does not have strongly differentiated cilia around it **Gymnostomina, 9**

**7b** Mouth has specially differentiated cilia around it, sometimes appearing as a membrane **8**

**8a** Mouth lined with rows of free cilia (F.4)
**Trichostomina**
(*Paramecium, Colpoda, Bresslaua, Tillina*)

**8b** Mouth surrounded by undulating membranes of fused cilia (F.5) **Hymenostomina**
(*Frontonia, Leucophrys, Tetrahymena, Glaucoma, Colpidium, Loxocephalus, Cinetochilum, Urocentrum, Uronema, Cyclidium, Pleuronema*)

**9a** Mouth at or near the front end on the upper side
**Protostomata, 10**

**9b** Mouth at the side, narrow or round
**Pleurostomata, 12**

**9c** Mouth on the underside, towards the front
**Hypostomata** (*Nassula, Chilodonella*)

**10a** Mouth has short spines (trichites) around it (F.6) and sometimes extensible projections used for entangling prey **Spathidiidae** (*Spathidium, Legendrea*)

**10b** Mouth without spines not at the tip of an apical projection looking like a cone **11**

**10c** Mouth without spines, at the tip of a cone
**Didineae** (*Didinium, Mesodinium*)

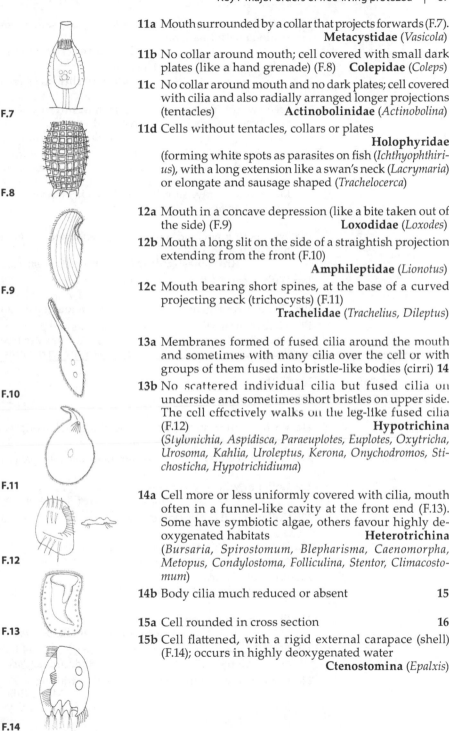

F.7

F.8

F.9

F.10

F.11

F.12

F.13

F.14

**11a** Mouth surrounded by a collar that projects forwards (F.7).
**Metacystidae** (*Vasicola*)

**11b** No collar around mouth; cell covered with small dark plates (like a hand grenade) (F.8) **Colepidae** (*Coleps*)

**11c** No collar around mouth and no dark plates; cell covered with cilia and also radially arranged longer projections (tentacles) **Actinobolinidae** (*Actinobolina*)

**11d** Cells without tentacles, collars or plates
**Holophyridae**
(forming white spots as parasites on fish (*Ichthyophthirius*), with a long extension like a swan's neck (*Lacrymaria*) or elongate and sausage shaped (*Trachelocerca*)

**12a** Mouth in a concave depression (like a bite taken out of the side) (F.9) **Loxodidae** (*Loxodes*)

**12b** Mouth a long slit on the side of a straightish projection extending from the front (F.10)
**Amphileptidae** (*Lionotus*)

**12c** Mouth bearing short spines, at the base of a curved projecting neck (trichocysts) (F.11)
**Trachelidae** (*Trachelius, Dileptus*)

**13a** Membranes formed of fused cilia around the mouth and sometimes with many cilia over the cell or with groups of them fused into bristle-like bodies (cirri) **14**

**13b** No scattered individual cilia but fused cilia on underside and sometimes short bristles on upper side. The cell effectively walks on the leg-like fused cilia (F.12) **Hypotrichina**
(*Stylonichia, Aspidisca, Paraeuplotes, Euplotes, Oxytricha, Urosoma, Kahlia, Uroleptus, Kerona, Onychodromos, Stichosticha, Hypotrichidiuma*)

**14a** Cell more or less uniformly covered with cilia, mouth often in a funnel-like cavity at the front end (F.13). Some have symbiotic algae, others favour highly deoxygenated habitats **Heterotrichina**
(*Bursaria, Spirostomum, Blepharisma, Caenomorpha, Metopus, Condylostoma, Folliculina, Stentor, Climacostomum*)

**14b** Body cilia much reduced or absent **15**

**15a** Cell rounded in cross section **16**

**15b** Cell flattened, with a rigid external carapace (shell) (F.14); occurs in highly deoxygenated water
**Ctenostomina** (*Epalxis*)

**F.15**

lorica
a vase-shaped structure
that surrounds a cell

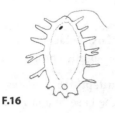

**F.16**

**F.17**

**F.18**

**16a** Cells with an external gelatinous lorica (F.15), sometimes studded with fine sand grains.
**Tintinnina** (*Tintinnopsis*)

**16b** Cells without an external lorica
**Oligotrichina** (*Halteria*)

**17a** Free-swimming but prone to attach to surfaces of invertebrates by hooks on the rear end
**Mobilina** (*Trichodina*)

**17b** Attaching normally to surfaces, sometimes with a lorica, sometimes with a stalk **Sessilina** (*Platycola, Cothurnia* (on gills of crayfish), *Vorticella, Epistylis* (on crayfish gills, turtle backs, chironomid larvae), *Carchesium, Zoothamnium*)

**18a** With both amoeboid movement (pseudopodia) and a flagellum (F.16) or flagella
**Rhizomastigida** (*Mastigamoeba, Cercomonas*)

**18b** With flagella but not amoeboid movement **19**

**19a** One or two flagella **Protomastigida, 20**

**19b** More than two flagella. Mostly symbionts and parasites (including *Giardia* and *Trichomonas*), but a few are free living **Polymastigida** (*Tetramitus*)

**20a** With one flagellum **21**
**20b** With two flagella **24**

**21a** With a collar surrounding the flagellum (collared flagellates, choanoflagellates) **22**

**21b** Without a collar **Oikomonadidae** (*Oikomonas*)

**22a** Collar or whole cell enclosed in jelly (F.17)
**Phalansteriidae** (*Phalansterium*)

**22b** Collar or cell not enclosed in jelly **23**

**23a** With a lorica (a vase-shaped structure surrounding the cell) **Bicosoecidae** (*Polyoeca, Bicosoeca*)

**23b** Without a lorica
**Codosigidae** (*Codosiga, Monosiga, Protospongia*)

**24a** Flagella equally long
**Amphimonadidae** (*Amphimonas*)

**24b** Flagella unequal in length, one of them trailing behind as the cell moves (F.18) **Bodonidae** (*Bodo*)

**24c** Flagella unequal in length with neither trailing
**Monadidae**
(can be solitary (*Monas*) or colonial (*Cephalothamnium, Anthophysis*))

**F.19**

**F.20**

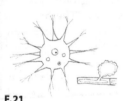

**F.21**

**25a** With numerous, stiff, unbranched radiating projections (F.19) emerging from the cell

**Actinopodia, Heliozoida**

(the marine equivalents are called radiolarians (*Actinosphaerium, Actinophrys, Radiophrys, Acanthocystis, Clathrulina*))

**25b** Without such projections      **Rhizopodea, 26**

**26a** With a shell (testa), usually single chambered, not of calcium carbonate (it will not dissolve in a drop of vinegar), (F.20) but may be studded with sand grains

**Testacida, 28**

**26b** Without a shell      **27**

**27a** With fine radiating, sometimes branched pseudopodia (extensions of the cell surface as it moves) (F.21); sometimes invading algal cells or filaments

**Proteomyxida** (*Actinocoma, Vampyrella*)

**27b** With blunt pseudopodia      **Amoebida, 31**

**28a** Shell simple and membranous      **29**

**28b** Shell of plates or studded with sand grains      **30**

**29a** Fine pseudopodia that form a network

**Gromidae** (*Gromia*)

**29b** Blunter pseudopodia that do not form a network

**Arcellidae** (*Arcella*)

**30a** Testa studded with sand grains or other particles

**Difflugidae** (*Difflugia*)

**30b** Testa of small plates of silica    **Euglyphidae** (*Euglypha*)

**31a** With an outer layer, or pellicle, that is stiff and may be thrown into ridges or folds as the cell moves

**Thecamoebidae** (*Thecamoeba*)

**31b** With a more fluid margin and movement      **32**

**32a** Movement on a broad front, the pseudopodia not well differentiated      **Amoebidae, 33**

**32b** Well defined pseudopodia, produced in a forward direction into which the cell contents can be seen to flow      **Mayorellidae**

(*Mayorella, Astramoeba, Flabellula*)

**33a** Movement by wave-like expansions

(*Pelomyxa, Valkampfia, Trichamoeba*)

**33b** Pseudopodia form randomly in different directions

(*Amoeba, Chaos, Metachaos, Polychaos*)

### 3.9 Invertebrates

Microscopic eukaryotes offer great scope for investigations on ponds, but require a compound microscope. Macroinvertebrates (those bigger than 1 mm or so) are equally valuable and much can be done with a good hand lens, though it is more comfortable and rewarding to use a stereomicroscope, which also gives access to the smaller invertebrates like rotifers, tardigrades and nematodes. Invertebrates are justifiably popular among amateur naturalists and often such people become national authorities on particular groups. Taxonomy and identification have become less popular among professional scientists as new problems use different approaches, but the need for accurate identification remains. There is plenty of scope for specialisation as older people die or retire and are not easily replaced. Invertebrates are well represented in freshwaters and there are more than 4,000 species in Britain and Ireland. Some groups, like the nematodes, are severely under-researched. The first task is to obtain samples and this is relatively easy unless the study requires minimal disturbance to the habitat, when it becomes more tricky.

Table 3.1 is a survey of all the groups of freshwater invertebrates in Britain and Ireland, the number of species, and some indication of the main ways of feeding. There are some groups, notably the crabs and prawns, which do not occur naturally here, but are very prominent in the tropics. Nonetheless, although the species and often genera will certainly differ, and our islands are relatively depauperate because of the effects of the recent glaciation (fifteen thousand years is very short in terms of establishment of stable communities), many groups and families are common everywhere. Indeed a good slice of the invertebrate scope, from sponges to arthropods is to be found in freshwaters, and although the diversity is commonly believed to be low compared with the ocean, the diversity relative to the extensiveness of habitat is exceptionally high. Only 0.02% of Earth's water is in surface freshwaters, whereas 97.6% is in the ocean, whereas about 10% of all known animal species and one third of vertebrates occur in freshwaters. The absolute difference in diversity depends probably on a matter of time and permanence that has allowed accumulation in the ocean, compared with the ephemerality afforded to freshwaters by volcanic action, glaciation and drought.

The ephemeral nature of freshwaters has been important in determining the nature of the organisms that grow in them. On a long time-scale, it has meant that there have

**Fig. 3.10** Many freshwater animals have evolved directly from marine ancestors (others via land intermediaries). Freshwater habitats are less predictable than marine ones and this is reflected in a generally drabber appearance of the animals. Freshwater animals must devote more energy to reproduction in a fluctuating habitat that may change rapidly and kill many individuals, whilst marine animals can afford to devote more energy to display and warning behaviour and colouration. Top to bottom: the freshwater crayfish; the European lobster; a common freshwater snail, *Bythinia tentaculata*; and a much more exotic looking marine snail. Photographs by David Gerke, Bart Braun and Lars Peters.

been many extinctions and reinvasions of organisms from the land, or, via the estuaries, from the oceans. There are virtually no specifically freshwater orders (just three possible groups of microorganisms) compared with 56 shared with the land, the ocean or both. Geologically, the freshwater biota is relatively young and there are striking resemblances in form with sister groups in the ocean (compare lobsters and crayfish (Fig. 3.10), marine and freshwater snails, bivalves, shrimps, and fish, for example). The land has been a major source for recolonisation and responsible for many recent entries, reflected in the aerial flowering of aquatic plants and aerial reproduction of many freshwater insects. On shorter time-scales, the ephemerality brought about by the risk of drought has meant that many invertebrates produce resting stages such as thick-walled structures that can survive weeks of drying and even baking heat, and most are drab in colour. This contrasts with the ocean, where virtually none of the animals produce resting stages, and where brightly coloured (and often poisonous) animals are common (Fig. 3.10). The risks of drying in freshwaters have meant that energy is devoted to reproduction rather than production of devices like warning colours or toxins to deter predators, as in the ocean. If a freshwater aquarium looks less exotic than a marine one, or a pond less than a coral reef, there is good reason.

**Table 3.1** Freshwater invertebrate fauna of Britain and Ireland. Feeding guilds are P, predators; Pa, parasites; S, leaf shredders; Sc scrapers F, filter feeders; Su, suspension feeders; and D, deposit feeders. Modified from Moss (2015).

| Group | Common name & main feeding guilds | Genera | Species | Notes |
|---|---|---|---|---|
| Porifera | Sponges: Su | 4 | 5 | Animals that have different sorts of cells but which are not organised into distinctive tissues. They pump water into an inner cavity and filter particles from it. |
| Cnidaria | Hydras, corals and jellyfishes: P | 4 | 7 | Rather like sponges except that they have organised tissues and catch prey by stinging it with special cells. Some forms have a polyp (attached to a surface) stage and a medusa (jellyfish) stage. Most British species have only the polyp stage but there is one (introduced species) where the jellyfish predominates. |
| Platyhelminthes | Flatworms: micro-turbellarians and triclads: P | 33 | 67 | As well as the microturbellarians and triclads, which are free-living predators there are many more parasitic tapeworms (cestodes) and flukes (trematodes), which infect vertebrates, often with a mollusc as an intermediate host. These are not listed here. The triclads are larger than the microturbellarians, which are tiny (1–2 mm) flatworms that attack other small animals. Triclads attack larger invertebrates by inserting the fore part of their gut into them and dissolving and sucking out the contents. All flatworms have a simple gut, open at only one end. Two of the 11 British triclads are essentially stream dwellers, but occasionally are found in cold northern lakes. Only one of 56 microturbellarians is confined to streams. |
| Nemertea (Rhyncocoela) | Ribbon or proboscis worms: D,Pa | 1 | 2 | There is a long proboscis hidden in a cavity when not used for feeding. Slow moving worms, rather little known. Mostly marine. |
| Nematomorpha | Gordian worms:Pa | 3 | 4 | Parasites of arthropods (of land and freshwater) with free-living adults that live in freshwaters. The adult worm is quite stiff and inflexible, sometimes with a hard 'plastic' feel. Little is known about their ecology. |
| Nematoda | Hair worms: Pa, P, D | 31 | 70 | Very abundant but under-recorded. Some are free-living, many are parasitic. |

**Table 3.1** *Continued.*

| Group | Common name & main feeding guilds | Genera | Species | Notes |
|---|---|---|---|---|
| Gastrotricha | Gastrotrichs or hairy backs: Sc, D | 11 | ?50 | Very small (<1 mm), covered in minute bristles. Probably detritus feeders on fine material. Under-recorded and often overlooked. |
| Rotifera | Rotifers: P, Su | 102 | 466 | Small animals (<1mm) but very diverse and widespread in both the plankton and on underwater surfaces. All have a ring of cilia around their mouths, which, in many, swirls a water current with suspended particles into the gut. Now includes the parasitic Acanthocephala (thorny headed worms). Some are capable of grasping larger prey. |
| Ectoprocta (Bryozoa) | Bryozoans, moss animalcules: F | 7 | 11 | Small colonial animals that attach to underwater surfaces and filter small particles from the water. Probably commoner in rivers than in lakes. |
| Mollusca [Gastropoda] | Snails, freshwater limpets: Sc,D | 31 | 46 | In freshwaters all have calcareous shells and feed by scraping material from surfaces. Some have cavities lined with blood vessels that allows them to breathe air and makes them amphibious. |
| Mollusca [Bivalvia] | Clams and mussels (swan, pea, pearl, zebra): Su | 10 | 32 | Some are attached to surfaces under water by threads and filter plankton from water pumped into them through a siphon. Others (generally bigger) are free-living in sediment and pump in a suspension of mud on which to feed. The Unionidae has larval stages that attach to fish as parasites. The freshwater pearl mussel is confined to rivers. |
| Annelida | Segmented worms, oligochaetes, aeolosomatids: D | 58 | ?23 | The freshwater equivalent of earthworms, though usually much smaller and finer. Sediment feeders, often pinkish in colour because of the haemoglobin many contain. |
| Annelida [Hirudinea] | Leeches: P | 12 | 16 | Only two are bloodsucking parasites (on fish or mammals). The remainder feed on invertebrates. |
| Tardigrada | Water bears: Sc,D | 4 | 42 | Very small (1–3 mm) fleshy animals that look like miniature bears and are common on wet mats of vegetation, particularly of *Sphagnum* at the edges of bog lakes. |

**Table 3.1** *Continued.*

| Group | Common name & main feeding guilds | Genera | Species | Notes |
|---|---|---|---|---|
| Arachnida [Hydracarina] | Water mites: P | 73 | 322 | Tiny (1–3 mm), often brightly coloured animals with the characteristic eight legs of arachnids but no differentiation of the body, as in spiders. The nymphs are parasitic on insects, the adults are predators. |
| Arachnida [Araneae] | Spiders: P | 1 | 1 | Just one species is truly aquatic, though many live on wet vegetation mats. The pond spider lives in a 'diving bell' of silk, full of air, and attached to water plants, where it awaits its prey. |
| Arachnida [Oribatei] | Oribatid mites: D | 3 | 4 | Most common in bogs on *Sphagnum* mats. |
| Crustacea [Anostraca] | Fairy shrimps: Sc,D,F | 2 | 2 | Live in temporary pools. Up to 35 mm, with a carapace and eyes on short stalks. Delicate-looking because of the feathery gills borne along both sides of the body. |
| Crustacea [Entomostraca] | Tadpole shrimp; triops: Sc,D | 1 | 1 | A rare animal of great antiquity in evolutionary terms and living in temporary pools. It has a rigid carapace but its eyes are not on stalks. |
| Crustacea [Cladocera] | Water fleas: F,Sc,P | 40 | 90 | Four distinct orders that are not closely related, but the term Cladocera has a long pedigree among freshwater ecologists. The common characteristic is a carapace that surrounds the animal and gives it a smooth outline. Planktonic or associated with vegetation and mostly filter feeders or nibblers on periphyton, though some are predators. Commonly parthenogenetic, producing females without the help of males, which appear only rarely. |
| Crustacea [Ostracoda] | Ostracods: F | 29 | 89 | Surrounded by a calcareous carapace of two halves joined at a hinge so that they look a little like swimming bivalves (though jointed legs protrude). Usually small (<3 mm) and filter feeding. |
| Crustacea [Copepoda] | Harpacticoid, calanoid and cyclopoid copepods: F,P,Pa | 51 | 118 | Small- to moderate-sized animals (up to 5 mm) with a streamlined body and no carapace, a pair of antennae used for rapid swimming (cf the slow movement of most Cladocera). Filter feeding, or predatory using grasping limbs. The females of some species are parasites on fish. |

**Table 3.1** *Continued.*

| Group | Common name & main feeding guilds | Genera | Species | Notes |
|---|---|---|---|---|
| Crustacea [Branchiura] | Fish lice: P | 1 | 3 | External parasites on fish, to which they attach by suckers that allow detachment and free-swimming. |
| Crustacea [Bathynellacea] | No common name, but ancient crustaceans: Sc,D | 1 | 1 | Interstitial in ground water. |
| Crustacea [Decapoda] | Crabs, shrimps, crayfish: Sc,D | 6 | 7 | Only one species is native, the white-clawed crayfish. The Chinese mitten crab only penetrates to lowland rivers and does not breed in freshwaters. The remaining five species are all crayfish introduced from mainland Europe and North America. |
| Crustacea [Mysidacea] | Mysids, opossum shrimps: Sc,D,P | 3 | 3 | *Mysis salemaai* (formerly *M. relicta*) is a plankton predator, supposed to be a glacial relict and was found in Ennerdale water (though not recently seen) and occurs in several Irish lakes. A second species, *Neomysis integer* is native to brackish waters, including some of the Norfolk Broads. A third species, the bloody-red shrimp, is introduced. |
| Crustacea [Isopoda] | Water hoglouse: Sc,D | 3 | 4 | The equivalents of woodlice on land and flattened top to bottom. |
| Crustacea [Amphipoda] | Freshwater shrimps: Sc,D,P | 8 | 21 | Body flattened from side to side. Several introduced from Europe or North America and three species are dwellers in caves and ground water, leaving four *Gammarus* species as native to lakes. |
| Myriapoda | Centipedes: D | 1 | 1 | Doubtfully a lake organism but may occur on marginal vegetation mats. |
| Hexapoda [Collembola] | Springtails: Sc,D | 13 | 22 | So called because of a forked appendage at the rear, which allows them to jump vigorously. Mostly semi-aquatic on the surface tension film or on vegetation mats, small (<5 mm). |

**Table 3.1** *Continued.*

| Group | Common name & main feeding guilds | Genera | Species | Notes |
|---|---|---|---|---|
| Hexapoda [Insecta] (Ephemeroptera) | Mayflies: Sc,D | 20 | 50 | About 18 in lakes, but none are confined to standing waters. Adults are aerial, nymphs are aquatic, with three fine 'tails' at the end of the abdomen. Gills are born on the abdominal segments. Some live on or under stones, scraping the films of organisms that grow there. Some burrow in sediments. None are predators. |
| Hexapoda [Insecta] (Plecoptera) | Stoneflies: Sc,P | 17 | 41 | About 7 in lakes, on stony shores, but none are confined to standing waters. More characteristic of streams. |
| Hexapoda [Insecta] (Odonata) | Damselflies and dragonflies: P | 20 | 44 | About 33 breed in lakes, with fewer than five requiring flowing waters. Damselflies (Zygoptera) are more slender than dragonflies (Anisoptera) both as larvae (aquatic) and adults (aerial). The abdomen of damselflies ends in three flattened gills. That of dragonflies in shorter, sharper spikes. Dragonfly adults hold their wings out at rest whilst damselfly adults fold them over the body. There are a further 14 vagrant species that do not normally breed. |
| Hexapoda [Insecta] (Hemiptera) | Bugs: P,Sc,D | 24 | 67 | Mouth parts are modified into beaks or rostra for piercing and sucking out the prey. Juveniles resemble the adults Some retain the ability to fly. Most (55) can be found in standing waters. The group includes water measurers, corixids, water boatmen, pond skaters, water striders, and water scorpion. |
| Hexapoda [Insecta] (Hymenoptera) | Wasps, ichneumon wasps: Pa | 32 | 41 | Mostly parasitic wasps that lay their eggs in the eggs or bodies of aquatic insects, but which are not truly aquatic themselves. |
| Hexapoda [Insecta] (Coleoptera) | Beetles: P,D | 111 | 415 | Both adults and larvae are aquatic, with the larvae looking very different from the adults, which retain their wings, protected by hardened wing covers, or elytra. The mouthparts are adapted for grasping, biting and tearing. A few have aquatic larvae and aerial adults, a few others vice versa. |
| Hexapoda [Insecta] (Megaloptera) | Alderflies: P | 1 | 3 | Predators with appendages of unknown function protruding from the abdominal segments in the larvae. Adults are aerial. |

**Table 3.1** *Continued.*

| Group | Common name & main feeding guilds | Genera | Species | Notes |
|---|---|---|---|---|
| Hexapoda [Insecta] (Neuroptera) | Spongeflies, lace-wings: P,Sc,D | 3 | 4 | Spongeflies feed as larvae on freshwater sponges. Lacewings are detritus feeders as larvae. Adults of all are aerial. |
| Hexapoda [Insecta] (Trichoptera) | Caddisflies: Sc,P,F | 78 | 203 | The adults are aerial, the nymphs aquatic, and characterised by two clawed prolegs at the tip of the abdomen. About 162 in lakes. Larvae may form cases of small stones, sand grains, twigs or segments of leaves, or even the shells of small snails and are then usually scrapers of films from rocks, or may be caseless (47 species) and either predators or spinners of nets to catch fine particles in streams. |
| Hexapoda [Insecta] (Lepidoptera) | China mark moths: Sc | 5 | 5 | The caterpillars are aquatic and make cases from leaf fragments. The adults are aerial. |
| Hexapoda [Insecta] (Diptera) | Two-winged flies: D, Sc,P | 357 | 1,610 | More probably occur in lakes than in rivers, but the group is surprisingly little investigated. Mostly the adults are aerial and the larvae aquatic and bearing little resemblance to the adult or even the familiar insect plan of head, thorax, abdomen and six legs. Head of larvae may be obscure, but there is usually a prominent spiracle for breathing on the final abdominal segment. There may be many yet to be described and some may be commoner in small puddles and water filled holes than lakes. The group includes mosquitoes, horse flies, tabanids, phantom midges, chaoborids, simuliids, ceratopogonids, biting midges, non-biting midges, chironomids, craneflies, moth flies, owl midges, soldier flies, daddy long legs, hoverflies, rat tailed maggots, and meniscus midges. |
| **Total** | | **1,326** | **4,127** | **Approximately 80–90% may be found in standing waters although many of these only on occasion or in very limited areas.** |

For the sampling of macroinvertebrates, the basic tool is a long-handled pond net. The net can be swept through the water in a bed of plants, and animals will be dislodged into it. Some that are firmly attached to surfaces, like freshwater limpets and bryozoans, may be underrepresented, as will those that burrow in the sediments, and those that are smaller than the mesh size of the net (usually 0.5 mm or in less expensive nets, 1 mm). In streams the procedure is to face downstream, hold the net firmly onto the bottom, and working upstream, disturb the bottom with the feet so that the animals are swept by the current into the net. This is called kick sampling but does not work so well in ponds because there is no current, and the net will fill rapidly with mud. It is better to sweep the vegetation and to take separate samples from the bottom mud using a corer. Suitable corers can be made from wide (say 8 cm) plastic piping with a rubber stopper fitting onto the lower end and threaded onto a wire or string that passes up through the tube. The stopper is kept to one side as the corer is pushed into the sediment to about 20 cm, then the string is pulled and the stopper blocks the tube and the corer is withdrawn. In stiff sediments, the mud may stay in the tube without a stopper, especially if the tube is full of water and sediment and a lid is fitted over the upper end. If a narrower corer (2 cm) is used looser sediment will stay in, at least long enough for the sample to be transferred to a bucket or jar.

However sampled, the animals need to be thoroughly washed through a net to remove fine sediments, and if gravel is present it needs to be picked out from the net. If it is not, the jostling of the sample on the journey home might pulverise the animals into a paste. The animals should be taken home in a jar with a lot of water. This might sound elementary, but sometimes people forget to add any water. It also helps to separate potential predators, like dragonfly and damselfly nymphs, and large water beetles and water bugs, into separate jars, or you may again find a much-depleted sample on return. It is then usual to pick out the animals into separate dishes or specimen tubes for identification. A small amount of the sample is placed in a white tray (butchers' display trays are ideal and cheap. Similar trays from a scientific supplier will be no better and much more expensive). Teaspoons, artists' paint brushes, pasteur pipettes and plastic forceps may all prove useful in retrieving animals, which can be fast moving. Vigorous stirring is not helpful. Animals will stay hidden within plant or other debris, but will emerge under quiet

**aliquot**
a portion of liquid

conditions. When you have exhausted the sample, pour away the remains and add another portion of unsorted sample. It is best to sort many small aliquots than to pour the lot into the tray in a vain attempt to speed things up. You will miss many animals if you do that. Animals sorted live can be returned to the pond, but if you wish to make a reference collection, 90% methylated spirits is a useful preservative though soft animals will contract. Labels are usually written on card in waterproof black ink and inserted into the specimen tube, which needs to be firmly stoppered to minimise evaporation of the spirit.

Many of the aquatic insects are aquatic only for the juvenile phases of their life history and emerge as aerial adults to mate and colonise other habitats. The larvae or nymphs of many species are readily identified but this may not be the case for many two-winged flies, especially the ubiquitous chironomids, which can only be identified as adults. Cages of soft insect netting can be constructed on wire frameworks mounted on floating wooden frames to catch these as they emerge, and commercial versions can be bought. Such cages are used in quantitative investigations where numbers per unit area are required or where studies are being made of the timing of emergence. For other purposes a standard wide insect sweep net can be used. It needs to be kept dry so some caution is needed lest overenthusiastic sweeping lands you and the net in the pond. Adult insects can be mounted on pins if you wish to keep a reference collection. Methods are given in Wheater & Cook (2003).

There are many specialist keys available but the initial step is to recognise the group to which an individual belongs. A brief examination with a hand lens, if necessary, at the pond side can tell you a lot if the major groups are recognised. Key G works in such circumstances. Once you are back indoors and the sample is sorted, your initial assessment can be confirmed with the general key developed by the Freshwater Biological Association (Dobson et al. 2012), which also includes some minority groups not included in Key G. The FBA key sometimes takes you to details of particular species but more often leads you to a more specialist key and gives details of what is available. The web sites of the Freshwater Biological Association and the Field Studies Council sell excellent specialist keys. Other useful general keys are given in the Bibliography (Chapter 10). Many freshwater studies, especially those concerned with processes, use only the family level for identification.

Beyond that the most common organisms are well known and recognisable to species, but concentration on families and common organisms is due to the fact that some groups are quite difficult unless one develops a particular interest and expertise in them. Again, the internet is a valuable resource with many images that are readily available to support your identifications.

## Key G Freshwater invertebrate groups (and a few genera) most commonly encountered

**1a** With a hard calcareous shell but without jointed limbs     **2**

**1b** Without a shell     **6**

**2a** Shells in one piece usually helical, but sometimes just pyramidal     **3**

**2b** Shell in two pieces hinged at one side     **5**

**3a** Shell closable by a flat plate (operculum) that can be moved across the opening and where the body can emerge (air breathers) or shell conical and not obviously helical     **4**

**3b** Shell with a prominent spire without an operculum (gill breathers), eyes on the outer sides of the tentacles
    **Freshwater winkles** (*Viviparus*)

**4a** Shell with a prominent spire and right handed (opening to right relative to spire when viewed from the side with the opening)     **Pond snails** (*Lymnaea*)

**4b** Shell with a prominent spire and left-handed
    **Bladder snails** (*Physa*)

**4c** Shell without a prominent spire, but constructed as flat coils     **Ramshorn snails** (*Planorbis*)

**4d** Shell conical sometimes with a slight hook at the apex     **Freshwater limpets** (*Acroloxus*)

**5a** Shell large (up to 10 cm)
    **Freshwater mussels** (*Anodonta, Dreissena*)

**5b** Shell small (usually <1 cm)
    **Pea** and **orb shells** (*Pisidium, Sphaerium*)

**6a** Body not divided into segments along its length     **7**

**6b** Body or sections of it obviously divided into segments     **10**

**7a** Organism bilaterally symmetrical, much longer than broad     **8**

**7b** Organism with radial symmetry or no apparent symmetry **9**

**8a** Thin hair-like worm, usually vigorously wriggling, colourless barely visible to the naked eye **Nematodes**

**8b** Flat worm, a few millimetres long, gliding over surfaces, with a triangular head and a projection (the pharynx) extending from the centre on the underside. White or brown. May be found attacking other macroinvertebrates **Flatworms** (triclads, planarians)

**8c** Flat worm, very small (<2 mm), planktonic, attacking zooplankters **Microturbellarians**

**zooplankter**
individual animal
member of the plankton

**8d** Small animal, up to 200 μm, with a circle of cilia around the mouth which beat to bring in water with suspended particles towards the mouth **Rotifers** (Rotifera)

**cilium (plural cilia)**
hair-like structure that
projects from the surface
of cells and can move

**9a** Sedentary organism without obvious symmetry on plants or debris, grey, brown or green with a spongy consistency **Sponges**

**9b** Radially symmetrical animal attached to surfaces, with a ring of tentacles surrounding the mouth (hydras) or suspended in the water with trailing tentacles (jellyfish) **Cnidaria** (formerly called Coelenterata)

**10a** Jointed limbs absent. Fleshy legs may be present **11**
**10b** Jointed limbs or appendages present **13**

**11a** No legs, no antennae **12**
**11b** Fleshy legs or antennae present, colourless, greenish, red or pink with distinct head in most cases, though hidden in some. **Two-winged fly larvae** (Diptera)

**11c** Small (a few millimetres but often smaller and barely visible to the naked eye) brownish or greyish animal with a head and eight fleshy legs **Water bears** (Tardigrada)

**12a** Thin, <2 mm often pink or red worm with bristles emerging from the segments, though these may only be visible under a microscope. Usually burrowing in sediment **Oligochaetes**

**12b** Thicker, brown or greenish worm with suckers at both ends **Leeches**

**13a** Three pairs of jointed limbs **20**
**13b** More than three pairs of jointed limbs **14**

**14a** Four pairs of jointed limbs **15**
**14b** More than four pairs of jointed limbs **16**

**15a** Four pairs of legs, small (a few millimetres), often brightly coloured and fast moving
**Water mites** (hydracarinids)

**15b** Four pairs of legs, larger (up to 1 cm body size and constructing a silken bell under water, or even larger and free moving over wet surfaces)     **Water spiders**

**carapace**
a cloak of exoskeleton that gives a smooth outline to the animal and may help in reducing drag

**16a** Hard carapace with strong pincers
**Crustacea** (crayfish)

**16b** Exoskeleton firm but not strongly calcareous and hard     **17**

**17a** Small motile organism usually much less than 4 mm likely to be found swimming in open water or over the sediments or around plants     **18**

**17b** Larger organism, usually moving over surfaces, up to 15 mm as adult     **19**

**18a** Body enclosed in a carapace which gives a smooth but not streamlined outline. Moving through the water with jumps as the antennae row through it. Eggs, if present, borne in a sac at the back of the carapace
**Crustacea** (water fleas, Cladocera)

**18b** Body streamlined with long antennae. Eggs held in sacs at the tip of the abdomen
**Crustacea** (copepods, Copepoda)

**18c** Bean-like animal with paired shells
**Crustacea** (ostracods, Ostracoda)

**19a** Body flattened top to bottom, all legs similar
**Crustacea** (isopods, e.g. water slater, *Asellus*)

**19b** Body flattened from side to side. Legs of different lengths and forms
**Crustacea**
(amphipods, e.g. freshwater shrimps, *Gammarus*)

**20a** With wings on underwater stages     **21**

**20b** Without wings (though there may be undeveloped wing pads) on underwater stages     **24**

**21a** Wings complete but front wings formed of horny covers (elytra), generally well developed biting mouthparts
**Coleoptera** (adult beetles)

**21b** Front wings horny in upper half, membranous in lower     **Hemiptera** (water bug), **22**

**22a** Living under water 23

**22b** Living on the water surface, moving over surface film and feeding on prey fallen on to the surface
**Pond skaters, water measurers, water crickets**

**23a** Long flattened tube emerging from rear of abdomen
**Water scorpions** (*Nepa*)

**23b** No tail tube, swimming ventral side uppermost
**Back swimmers, water boatmen** (*Notonecta*)

**23c** No tail tube, swimming back uppermost **Corixids**

**24a** Free-living 26

**24b** Living in a case or tube 25

**25a** Living partly in a case made of sand grains, leaf fragments or woody debris, with two hooks at the end of the abdomen and able to move around taking the case with it **Cased caddis larvae**

**25b** Living in a tube cemented to a surface but able to move out of the tube. Usually with prominent gills on the abdomen and two hooks at the end of the abdomen
**Free-living caddis larvae**

**25c** Living in a tube, with five pairs of fleshy false legs on the abdomen **China mark moth caterpillars**

**26a** Free-living with one projection from the end of the abdomen and seven pairs of long thin gill projections on the abdomen
**Megaloptera** (alder fly larvae, *Sialis*)

**26b** Free-living with more than one projection from the tip of the abdomen 27

**27a** Two projections from the tip of the abdomen 28

**27b** Three projections from the tip of the abdomen 29

**28a** Free-living with generally two short and stubby projections at the end of the abdomen **Beetle larvae**

**28b** Free-living with two fine projections emerging from the end of the abdomen
**Plecoptera** (stonefly nymphs)

**29a** Three fine bristle-like projections
**Ephemeroptera** (mayfly nymphs)

**29b** Three long, paddle-like, flat plates
**Anisoptera** (damselfly nymphs)

**29c** Three short stubby projections
**Zygoptera** (dragonfly nymphs)

# 4 Plankton

One of the more frustrating experiences in amateur pond dipping can be to sample the plankton. If a bottom net is used, virtually nothing will be caught and if a drop of water is put under a microscope, there will usually be little to be seen. What is needed is a proper plankton net, with a mesh size of less than 60 μm (and even then some of the community will pass through) or an inverted microscope in which the organisms present in several millilitres of water are allowed to settle to the bottom in a special chamber before being examined from beneath. An inverted microscope is a very expensive item and plankton nets are more expensive than pond nets, but good nets are worth the investment for the plankton is a fascinating community (Fig. 4.1).

Plankton will develop in most waters provided the residence time is long enough and that generally means a week and usually more. It is a very specialist community, with passively drifting viruses, bacteria and algae (virioplankton, bacterioplankton and phytoplankton), and weak

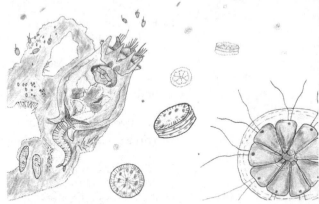

**heterotrophic**
obtains its own food from other organisms and does not photosynthesise

**raptorial**
grasping, as in raptors (hawks, falcons, eagles) among birds

**Fig. 4.1** Compared with the communities that form on plant surfaces and mud, the plankton is much less dense and the cells well-spaced. There are two sources of energy: organic matter washed in, like the irregular lump to the left, and photosynthesis carried out by the phytoplankton. Dead organic matter is colonised by bacteria, and a food chain forms with ciliates (lower left), heterotrophic nanoflagellates, the small flagellated cells, and rotifers (*Brachionus plicatilis*, to the left). Only the bigger cladocerans and copepods can cope with the algal food that dominates here, because it is too big for ingestion by the rotifers. The circular diatom, *Cyclotella*, is widely edible, but the colonial green alga, *Pandorina*, can only be taken by raptorial copepods that can bite pieces from it.

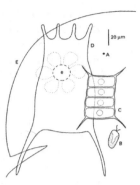

**Fig. 4.2** Relative sizes of some major components of the plankton. (A) a bacterium; (B) *Cryptomonas*, a relatively small phytoplankter; (C) *Scenedesmus*, a moderately large phytoplankter; (D) *Keratella*, a rotifer, a small zooplankter; (E) outline of the head with eye (e), of *Daphnia*, a large zooplankter. The head constitutes about a quarter to a fifth of the total body size.

swimmers among the flagellated algae and zooplankton; it is easily washed out of a lake or pond with vigorous flow-through. A net will not concentrate viruses or bacteria, nor the many very small algal cells, so will give a sample biased towards the larger algae and animals. It may also give a biased sample of the zooplankton because many of the animals can detect the movement of the net as it is pulled through the water and may move out of its way. Our usual way is thus not to pull the net but to take buckets of water, or use a special sampler that encloses a volume of water at depth and pour the sample (usually 50 litres or more) through the net. Even then some zooplankters will avoid the bucket.

What is obtained will be a concentrated sample, greenish or greyish, depending on the concentration of algae, which you can look at under a microscope. Stereomicroscopes are good for most zooplankters, compound microscopes desirable for rotifers, algae and bacteria. Of course the spacing of the organisms in the open water will be much greater than seen in the concentrated sample, so it is wise to have some appreciation of the real spacing to understand how the community functions and how it has solved the problems of persisting in a dilute, shifting and dangerous environment. To do this we can create an analogue in which you imagine you are the size of one of the smaller zooplankters, a rotifer (Fig. 4.2), at about 100 μm and then the other organisms are scaled up to objects familiar to you at your normal size.

In a typical pond water, rotifers might be present at densities of about 1,000 per litre and therefore occupy a volume of about 1 ml each. In the scaled-up model, the next rotifer to you will be about one hundred person-lengths, say 200 m, away. The viruses, with up to ten million per ml, will each occupy 100,000 $\mu m^3$, and the nearest ones will be very close, about one third of a millimetre distant, or in the scaled-up model, lentils about three person lengths or 6 m. Bacteria will number around a million per ml, so in the model, they will have the size of peanuts and be ten person lengths, 20 m, away from you. Bacteria are of the order of 0.5 to 2 $\mu m$ across and there is then a steady continuum through the smallest algae. Those that are 10–30 $\mu m$ across might be present at 1,000 cells per ml. In the model they are represented by plums, small bananas, apples or pears (and indeed they are often very different from one another), with a spacing of over a millimetre in reality, 10 person lengths or 20 m in the model. Bigger algae may often be colonial with colonies up to 80 $\mu m$ across, at densities of 100 per ml and spacing of over 2,000 $\mu m$ (watermelons, large pumpkins, at distances of 40 m in the model. They may also be filamentous and represented in the model by scarves, or streamers of toilet roll, a couple of metres long. The biggest colonial algae may reach 1 mm across (large cows, heavy horses) and are thus bigger than rotifers. Densities might be 1 per ml so the spacing would be 10,000 $\mu m$ (1 cm) or 200 m in the model. The larger zooplankters, water fleas and copepods, can be from 0.5 mm or less to several mm long, so overlap in size with the largest algae. A very large copepod (a small elephant), 4 mm long, might be present at 10 per litre, thus spaced at 10 cm in reality or one thousand lengths, 2 km, in the model. If we envisage the smaller fish, growing up to 10 cm in their first year, the analogy would be with the largest whales and the spacing perhaps one metre apart in reality. For you as a model rotifer, the nearest one, on average, would be 20 km away.

From this you will get the idea of the plankton as a well spaced out community, compared with say the bottom mud or the surfaces of plants in the littoral zone, where the bacteria and algae jostle over one another, and the invertebrates are only centimetres distant from one another or even closer. These surfaces tend to be richer in nutrients but the grazers and predators are also denser. Rotifers also occur among and attached to the plants. It is useful to carry out this same exercise of scaling the components up to human size based on your own estimates of the population densities per unit area of bottom or plant surface. T.T. Macan did it very graphically for invertebrates in Hodson's Tarn in the English Lake District, in *Ponds and Lakes* (1973).

## 4.1 The importance of spacing

The spacing of the plankton carries a number of consequences. First it allows enough light to reach algal cells deeper in the water column. There is a limited amount of light under water. Concentrating cells in a surface layer would prevent any photosynthesis deeper down and rapidly deplete all the nutrients from the surface layer. A more even distribution allows exploitation of a much larger volume of water. Consequently most plankters do not float but are denser than water and steadily sink until picked up by a wind-driven eddy current and returned towards the surface. Only a few are less dense, largely cyanobacteria that have gas vesicles in their cells and tend to float upwards, but accumulation at the surface means exposure to ultraviolet light, and high levels of visible light, to which their pigments are susceptible. The same eddy currents that pick up sinking cells also entrain floating cells and colonies, move them downwards and maintain them in more favourable conditions. When an algal bloom forms as a scum at the surface (Fig. 4.3), it is usually because this mechanism has failed as a result of an overproduction of gas vesicles.

There is meaning also to the size range found in the bacterioplankton and phytoplankton, in our model from lentil to heavy-horse-sized. The smaller the cell, the lower

**Fig. 4.3** Cyanobacterial bloom on a temple tank in Nepal. Photograph by K.J. Irvine.

is its sinking rate (Chapter 2) and therefore the risk that it will reach the sediments and be entrapped in the mud, or eaten by the grazers present in greater density there. But the smaller the cell, the greater is the range of filter-feeding zooplankters that can take it, so its risk of being grazed in suspension increases. Large colonies have the advantage that they cannot be ingested by most planktonic animals and those that can take them tend to bite off lumps, leaving the rest of the colony to keep growing, but they sink rapidly. Smallness also means a large surface area in relation to volume, and therefore advantages in taking up scarce nutrients, but largeness means faster sinking and an ability, through that, to keep on reaching 'new' water that might be slightly richer in nutrients. Animals like fish, moving through the water, release plumes of excreted nutrients (like the contrails of vapour that follow aeroplanes) that lead to a patchiness in nutrient availability. Nutrients are naturally scarce and nutrients in such plumes are rapidly taken up by algae that are lucky enough to drift through them.

The diversity in size of the phytoplankton is great and diversity is increased even more by the different pigment systems that support chlorophyll in photosynthesis, and which are used in the keys to distinguish particular groups. These pigments can use different wavebands. Green pigments, like chlorophylls, absorb best in the red; yellow, brown and orange pigments (xanthophylls and carotenoids) in the blue; red pigments in the green; and blue pigments in the yellow. The cyanobacteria and red algae are rich in blue (phycocyanin) and red (phycoerythrin) bile pigments. Because the dominant colour (the 'light climate') changes with depth in the water, a range of pigment combinations allows efficient use of what light is available by a community continually swirling up and down the water column. A typical phytoplankton community might have twenty or so species relatively abundant and actively growing at a given time. But as the season changes, other species will become favoured and existing ones handicapped, so that waiting in the wings are some hundreds more species that simply tick over most of the time but surge into activity for a short period when conditions favour them (Fig. 4.4). We typically measure the overall phytoplankton community by extracting the pigment chlorophyll *a*, common to all of the eukaryotes and the cyanobacteria, and this is often useful but it gives us little insight into the complexity of the community and its many different organisms.

**bile pigment**
coloured compound formed as a metabolic product of substances called porphyrins

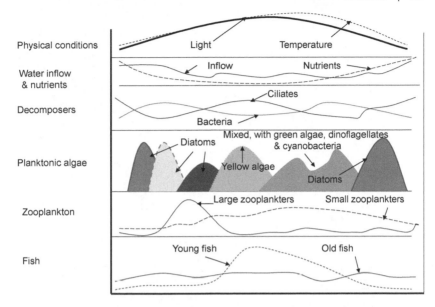

Physical conditions — Light    Temperature

Water inflow & nutrients — Inflow    Nutrients

Decomposers — Ciliates / Bacteria

Planktonic algae — Diatoms / Mixed, with green algae, dinoflagellates & cyanobacteria / Yellow algae / Diatoms

Zooplankton — Large zooplankters    Small zooplankters

Fish — Young fish    Old fish

Jan  Feb  Mar  Apr  May  Jun  Jul  Aug  Sep  Oct  Nov  Dec

**Fig. 4.4** Changes in the plankton community in a north-temperate pond. As daylength increases and water temperatures rise, conditions become more favourable for planktonic algal growth. Nutrients have built up over winter and fuel this. There is a fluctuating input of organic matter as temperature affects activity in the surrounding land and inflow of water brings in dissolved organic matter, on which bacteria and thence ciliates and other protozoans feed. There is a succession of many different species of planktonic algae, determined by changing circumstances of inflow and washout, nutrient availability, chemical secretion by the algae themselves, and zooplankton grazing. Meanwhile the zooplankton community is itself modified by fish predation following the hatch of young fish in early summer.

## 4.2 Zooplankters

Parallel variety in biology is found in the zooplankters. There are three main kinds and a handful of less common groups (Fig. 4.5). The main kinds are the rotifers, the cladocerans (water fleas) and the copepods, and a typical net sample will produce representatives of all of these. The less common groups are jellyfish (one species that has been introduced to Britain), microturbellarians and phantom midge larvae, all of which are predators on the main groups. You will also come across ciliates and colourless flagellates, which are indeed very abundant but some of which pass through the nets. These protozoans, along with bacteria and sometimes rotifers, tend to be associated with lumps of organic detritus, which are also commonly suspended

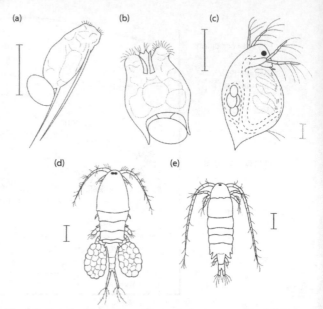

**Fig. 4.5** Zooplankters come from several groups but three are predominant. (a) *Filina* and (b) *Brachionus* are rotifers, on which the ring of cilia used in suspension feeding can be seen at the upper end, and parthenogenetic eggs at the lower end. (c) is a filter-feeding cladoceran, *Daphnia*, with a smooth carapace enclosing the filtering limbs and an egg pouch to the back. It swims by beats of the antennae. (d) and (e) are cyclopoid and calanoid copepods. Cyclopoid copepods are raptorial feeders and several species are common in ponds. They have a smoothly tapering body, short antennae and two egg sacs (on the female). Calanoid copepods are scarce in small water bodies but are filter feeders with long antennae, a more oblong body and one egg sac.

in the water. The 'purest' mountain tarn water is still a dilute suspension of organic junk, containing fish and zooplankton faeces. The planktonic food web has one set of components feeding on incoming and internally produced detritus (faeces, and algae killed by parasites, for example) and the other based on phytoplankton photosynthesising in situ. Filter-feeding zooplankters will graze more or less indiscriminately of the taxonomic origin of their food, on organisms involved in both of these pathways, but they do discriminate on size and sometimes on chemistry of the potential food.

The rotifers are the smallest of the three main groups, mostly around 100–200 μm in size. They have rings of cilia around their mouths, whose rhythmic beating creates a circular current that swirls water, bringing with it a stream of planktonic particles, into the mouth. Such suspension feeding is not very efficient as there is no concentrating mechanism and the size of the mouth confines rotifers to very small particles (about 1–5 μm) of bacteria and the smallest algae and protozoans. They do not compete very well therefore with other zooplankters that filter feed by concentrating particles that are bigger. The rotifers have a couple of major advantages though. They are small and unattractive to fish that must be able to see their prey to attack. Very small fish fry will take rotifers but this phase lasts only a short time. But rotifers' biggest advantage for survival is that they reproduce parthenogenetically. Females produce more females from eggs that need not be fertilised and can be produced, hatch and develop very rapidly. In a hostile environment (for the open water is dilute in food and the food is often of low nutritional value, there is little cover and the medium is continually churned by the wind) this gives a major advantage. Where death rates are high, reproductive rates must also be high to compensate. A few rotifers are raptorial, which means that they can grasp larger food and, for the larger ones, swallow it. The food might be a smaller rotifer or a large algal cell. Sometimes the cell is squeezed to release its contents, which are then swirled into the mouth.

**carapace**
a cloak of exoskeleton that gives a smooth outline to the animal and may help in reducing drag

The water fleas are often filter-feeders, with a carapace enfolding their bodies and the filtering limbs born on the undersides; there is an egg sac at the back. Beating of the limbs brings water past them and closely spaced hairs retain particles generally from about bacterial size (1 μm) to middle-range algae, perhaps 50 μm across. Bigger particles or long filaments are clawed out and rejected by specialised limbs, and energy must be used to do this, so feeding is less efficient where large particles dominate. Water fleas are also parthenogenetic. As in rotifers, males are only produced sporadically, often in response to adverse conditions when fertilisation occurs and resting eggs in protective cases are produced. This rapid production of parthenogenetic eggs also compensates for high death rates, because water fleas are very vulnerable to fish predation. They are larger than rotifers and move only slowly in the water by a series of short and jerky jumps and are easily seen. As with rotifers, some water fleas are raptorial, grasping smaller zooplankton prey and eating it, often after a struggle to manoeuvre the prey into a favourable position for swallowing.

The copepods are very different. They are streamlined, fast moving and adept at avoiding attacks by most fish by responding to the vibrations of sensory hairs on their antennae and moving rapidly out of the way. In many lakes on the European mainland, they (and also water fleas and rotifers) are vulnerable to specialist filter-feeding whitefish (coregonids) that cruise rapidly through the water, but such fish are very rare in the UK, occurring in only a handful of large lakes. Copepods can filter-feed on particles in a slightly larger size range (5–80 μm) than water fleas but less efficiently, and are slow reproducers, with a need for a sexual stage and then a succession of six stages of small filter feeders called nauplii and six further copepodite stages, the last of which is the adult stage that reproduces. The life history, egg to egg, takes several weeks compared with only a few days in the parthenogenetic zooplankters.

In surviving in the open water, therefore, both for phytoplankters and zooplankters there are many ways to swing the cat, and given a degree of structure of the water mass through stratification and the physical and chemical changes that occur during the year, creating the basis for many niches, it is not surprising that the overall community might include a couple of hundred phytoplankters and some tens of zooplankters, boosted in ponds often by small animals and algae swept up from the edges and bottom. There are readily available keys to the zooplankters that might be found in Britain and Ireland, but phytoplankters are less well served. Key H is to the most commonly encountered algal genera in the plankton of ponds and serves as a stepping stone to coping with the more specialist works if used in the same way as the keys give in Chapter 3. There are about 200 planktonic genera of algae in Britain and Ireland. This key lists about 80. However, the key will bring you into the region of similar organisms based on colour and overall structure. You should then be able to find the correct genus and species by looking at internet photographs and consulting the standard works (see Chapter 10, particularly John *et al.* 2011). But sometimes there is uncertainty even among expert taxonomists!

## Key H Common phytoplankton genera in ponds in Britain and Ireland

**H.1**

**1a** Cells swim, using flagella (fast or jerky active movement, or flagella visible)  **2**

**1b** Cells do not swim with flagella, but there may be a slow creep  **20**

**2a** Unicellular  **3**

**2b** Colonial  **12**

**3a** Cell covered with a case that is beyond the cell wall  **4**

**3b** Cell not covered by a case  **7**

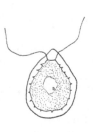

**H.2**

**4a** Case very obvious, yellow, brown or black, (H.1) sometimes obscuring the cell inside  **5**

**4b** Case transparent but may be wide, two flagella (H.2)  **6**

**5a** Case spherical or egg-shaped, sometimes with a short neck through which a single flagellum emerges. Case may be smooth or covered in short spines.
*Trachelomonas* (euglenoid)

**5b** Case flattened (lenticular), may be patterned, two flagella  *Phacotus* (green alga)

**H.3**

**6a** Cell outline lobed (H.3)  *Lobomonas* (green alga)

**6b** Outline smoothly oval, not winged, cells often with red pigment (carotene) masking the green. Often found in rainwater gutters  *Haematococcus* (green algae)

**6c** Case indented at the middle, wider at the circumference in side view, circular in front view
*Pteromonas* (green alga)

**7a** Chloroplast brownish, russet or yellowish  **8**

**7b** Chloroplast green  **9**

**7c** Chloroplast blue, cell drop-shaped
*Chroomonas* (cryptophyte)

**7d** Cell coloured purple, sausage-, cigar- or spirally-shaped, usually in deoxygenated habitats such as ponds with thick layers of fallen leaves
**Purple sulphur bacteria** (Fig. 5.9)

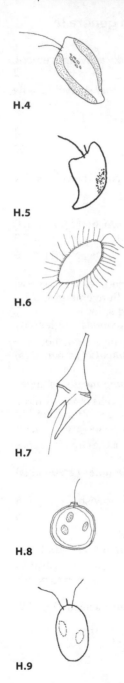

**H.4**

**H.5**

**H.6**

**H.7**

**H.8**

**H.9**

**8a** Cell large (>30 μm), drop- or heart-shaped, with a curve or coming to a blunt point and sometimes with a groove along one side (H.4)     *Cryptomonas* (cryptophyte)

**8b** Cell small (<15 μm), triangular and sharply pointed (H.5)     *Rhodomonas* (cryptophyte)

**8c** Cell lemon- or egg-shaped, often with one of more long spines at the end and having a rough appearance owing to the scales that cover the surface. The scales sometimes have many fine spines that project (H.6)
     *Mallomonas*
(yellow-green alga) (or cells detached from a colony of *Synura* (yellow-green alga))

**8d** Cell with a prominent groove around the middle, often with distinctive plates forming the wall (C.6)
     *Peridinium* (dinoflagellate)

**8e** Cell with long projections upwards and downwards and with groove around the middle. Common in large lakes, less common in ponds (H.7)
     *Ceratium* (dinoflagellate)

**8f** Cell with a prominent groove around the middle, but without obvious plates forming the wall
     *Gymnodinium* (dinoflagellate)

**8g** Cell irregularly rounded, delicate, chloroplast pale (a genus that has several feeding modes, including the engulfing of bacteria)
     *Ochromonas* (yellow-green alga)

**8h** Cell contained in a mineral covering (a lorica), with a pore from which the flagellum emerges (H.8)
     *Chrysococcus* (yellow-green alga)

**8i** Very small cell, with two flagella and a small projection (the haptonema) between them (H.9). In brackish waters, such as ponds or ditches influenced by sea water     *Prymnesium* (haptophytan)

**9a** Cells spherical or elongated but not flattened     **10**

**9b** Cells prominently flattened     **11**

**10a** Cells spherical or oval, two flagella
*Chlamydomonas* (green alga)

**10b** Cells bluntly elongated, can change shape, one flagellum emerges from cell *Euglena* (euglenoid)

**10c** Cells prominently spiralled, one flagellum
*Lepocinclis* (euglenoid)

**10d** Cell very long and thin, spindle-shaped and pointed. Two flagella emerge from one end
*Chlorogonium* (green alga)

**10e** Cell banana-shaped, two or four flagella, swims in spirals *Spermatozopsis* (green alga)

**10f** Cell strawberry-shaped with four lobes; four flagella emerge from a pit at the broad end, among the lobes (H.10) *Pyramimonas* (green alga)

**H.10**

**11a** Cell oval, four flagella emerge from a depression, no spiral markings *Platymonas* (green alga)

**11b** Cell round or oval, pointed, with prominent spiral or strip-like markings, one flagellum
*Phacus* (euglenoid)

**11c** Cell bean-shaped, two flagella emerge from indented side, swims sideways (H.11)
*Nephroselmis* (green alga)

**11d** Cell an oval slightly flattened on one side, in front view, indented in side view, with two flagella emerging from the indentation *Mesostigma* (green alga)

**12a** Motile colony, cells brownish or yellowish **13**

**12b** Cell green **14**

**H.11**

**13a** Cells densely packed into spherical or sometimes more elongate colonies that tend to break apart easily, releasing cells that are similar to *Mallomonas* (8c above) *Synura* (yellow-green alga)

**13b** Cells appearing at the outer edges of a colony mostly taken up by a large sphere of mucilage
*Uroglena* (yellow-green alga)

**13c** Cells densely packed into more or less spherical colonies and each bearing a spine emerging from a small projection on the surface of the cell
*Chrysosphaerella* (yellow-green alga)

**H.12**

**13d** Each cell contained in an open, vase-like structure, the vases linked together by threads into a branching colony H.12) *Dinobryon* (yellow-green alga)

**14a** Colony flat **15**

**14b** Colony more or less spherical **16**

**15a** Four or sixteen cells in a square arrangement
*Gonium* (green alga)
**15b** Sixteen cells in a horseshoe arrangement, with the colony curved at the forward end but flat at the rear
*Platydorina* (green alga)

**16a** Four cells in a small compact bunch
*Pascherina* (green alga)
**16b** More than four cells 17

**17a** Sixteen, thirty-two or sixty-four cells 18
**17b** Hundreds of cells, linked by fine threads
*Volvox* (green alga)

**18a** All cells the same size 19
**18b** Cells to the rear smaller than those to the fore
*Pleodorina* (green alga)

**19a** Cells packed solid in the centre of the colony (Fig. 4.1)
*Pandorina* (green alga)
**19b** Cells spherical occurring around the outside of a hollow mucilaginous sphere
*Eudorina* (green alga)
**19c** Eight large cells surrounded by mucilage, through which projections pass, the whole then held in a larger mucilage sphere *Stephanosphaera* (green alga)

**isodiametric**
roughly spherical, the same diameter from all angles

**20a** Single cells 21
**20b** Colonies (more or less isodiametric) 40
**20c** Filaments 55

**21a** Green 22
**21b** Brown or yellow (diatoms; walls of dead cells are readily preserved, being of silica, and will show fine patterning. If no dead cells are available, you will need cleaned material (see p.52) 36

**22a** Cell elongate, >5 times as long as broad 23
**22b** Cell not elongate, <5 times as long as broad 26

**23a** Cell straight 24
**23b** Cell distinctly curved 25

**H.13**

**H.14**

**H.15**

**24a** Cells single or in pairs with thick mucilaginous coating (H.13)                    *Elakatothrix* (green alga)

**24b** Cell without thick mucilaginous coat, spindle-shaped with long point at one end and two horizontally splayed processes at the other (H.14)          *Ankyra* (green alga)

**24c** Cell 10 to 15 times as long as broad with bluntly pointed ends
                    *Ankistrodesmus (Monoraphiphium)* (green alga)

**24d** Cell very thin, needle-like with sharp pointed ends
                    *Raphidonema* (green alga)

**25a** Cell large, moon-shaped, with several circular bodies (pyrenoids) located along the two chloroplasts, which often show longitudinal ridges. Line along middle divides cell into two halves (H.15)
                    *Closterium* (green alga, desmid)

**25b** Cell moon-shaped but very thin, single or in bunches, no pyrenoids          *Ankistrodesmus* (green alga)

**25c** Cells curved like a comma  *Selenastrum* (green alga)

**26a** Cell with spines                                               27
**26b** Cell without spines                                           31

**27a** Fewer than six spines per cell                          28
**27b** More than six spines per cell                          29

**28a** Spines short, sometimes barely perceptible, cells four- or five-angled          *Tetraëdron* (green alga)

**28b** Ellipsoidal cell, spines four, long, emerging at right angles to one another          *Lagerheimia* (green alga)

**29a** Spines in bunches emerging from the angles of an irregular cell          *Polyedriopsis* (green alga)

**29b** Ellipsoidal cells, spines emerging in bunches at the poles          *Chodatella* (green alga)

**29c** Cell ellipsoidal, spines delicate and emerging over the whole surface          *Siderocystopsis* (green alga)

**29d** Cell spherical, spines emerging over the whole surface                                               30

**30a** <8 μm diameter          *Golenkiniopsis* (green alga)
**30b** >8 μm diameter          *Golenkinia* (green alga)

**31a** Cell flattened                                               32
**31b** Cell globular or ovoid, not flattened                33

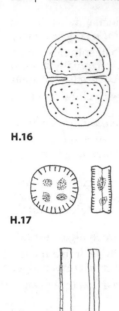

**H.16**

**32a** Cell divided in the middle by a deep groove into two identical halves. 'Cottage-loaf' shape (H.16)
*Cosmarium* (green alga, desmid)
**32b** Cell triangular, yellowish green rather than grass green *Goniochloris* (xanthophyte)
**32c** Cell four- or five-angled *Tetraëdron* (green alga)
**32d** Cell round, disc-like, yellow-green rather than grass green *Trachychloron* (xanthophyte)

**33a** Wall with brown spots or dots 34
**33b** Wall without brown spots or dots 35

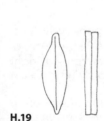

**H.17**

**34a** Cells irregularly ovoid with dots all over
*Siderocoelis* (green alga)
**34b** Cells regularly ovoid. Dots concentrated towards poles; young cells often retained within wall of mother
*Oocystis* (green alga)

**35a** Cell cylindrical, sometimes short filaments of two or three cells
*Stichococcus* (green alga)
**35b** Cell spherical or slightly ovoid *Chlorella* (green alga)
**35c** Cell ellipsoidal, lemon-, oval- or barrel-shaped
*Oocystis* (green alga)
**35d** Cell crescent-shaped *Ankistrodesmus* (green alga)
**35e** Cell short, spindle-shaped
*Scenedesmus* (single-celled form) (green alga)

**H.18**

**36a** Cell circular in front view, rectangular in side view (H.17) 37
**36b** Cell rectangular in both views (H.18) 38
**36c** Cell boat- or slipper-shaped in front view, rectangular in side view (H.19) 39

**H.19**

**37a** Cell-wall markings in two distinct concentric zones (H.20) *Cyclotella* (diatom)
**37b** Cell-wall markings uniform from centre to edge
*Stephanodiscus* (diatom)
Note that these two genera have been extensively revised in recent years, to include several more genera but identification on the new classification requires an electron microscope for confirmation.

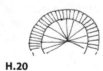

**H.20**

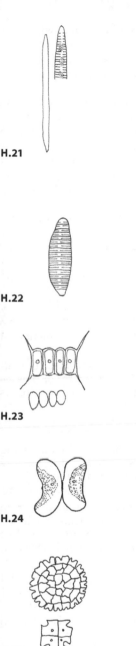

H.21

H.22

H.23

H.24

H.25

**38a** Two chloroplasts arranged fore and aft. Heavier teeth-like markings, often forming a ladder) at longitudinal edge. Motile with smooth gliding motion
*Nitzschia* (diatom, long, thin species)

**38b** More than two (usually many more) chloroplasts; more needle-like, but can be squatter; non-motile, markings uniform across the cell (H.21)    *Synedra* (diatom)

**39a** Two chloroplasts, arranged longitudinally, side by side    *Navicula* (diatom, but could also be several other genera of diatoms that live on sediment but are sometimes dislodged into the plankton. Cleaned material (see p.52) and reference to Key E will be needed for confirmation.)

**39b** Two chloroplasts arranged fore and aft
*Nitzschia* (diatom, short species)

**39c** More than two chloroplasts, more or less randomly arranged. Cell with heavy transverse silica partitions, visible in side (girdle) view (H.22)
*Diatoma* (diatom)

**40a** Bright grass-green    **41**

**40b** Brown, brown-green, olive-green, blue-green, black or pink but not grass-green    **47**

**41a** Colony flat    **42**

**41b** Colony a spherical or irregular mass but not flat    **44**

**42a** Colony a line (like a fence) of 2, 4, 8 or even 16 parallel cells, with or without spines (H.23)
*Scenedesmus* (green alga)

**42b** Colony of two small, sausage-shaped cells, without spines, joined at centre by a pad of mucilage (H.24)
*Didymogenes* (green alga)

**42c** Colony of four cells, arranged radially    **43**

**43a** Rounded cells, with spines    *Tetrastrum* (green alga)

**43b** Distinct squarish appearance, cells not spined
*Crucigenia* (green alga)

**43c** Colony like a plate of at least four, and generally many more, cells with those at the edge having projections or indentations. Within the plate, the cells may be very irregular in shape and there may be holes between them (H.25)    *Pediastrum* (green alga)

**44a** Cells packed together    **45**

**44b** Cells more loosely held, with space between them, sometimes strung with mucilage strands    **46**

**H.26**

**45a** Colony round, regular, outer cells sometimes with short projections, not storing oil    *Coelastrum* (green alga)

**45b** Colony an irregular mass that releases oil globules if squashed or pierced    *Botryococcus* (green alga)

**46a** Cells slightly ovoid, attached distantly by narrow strands of mucilage, like balloons on strings (H.26)
    *Dictyosphaerium* (green alga)

**H.27**

**46b** Cells spherical, attached side by side in irregular groupings, with one or two long spines emerging from each cell (H.27)    *Micractinium* (green alga)

**46c** Cells cigar-shaped, attached at one end and radiating out from the centre of the colony (H.28)
    *Actinastrum* (green alga)

**H.28**

**47a** Cells blue-green, pink or blackish, or pale with black dots in the cells    **48**

**47b** Cells brown, olive-green or brown-green    **52**

**48a** Colony made up of coiled filaments, parallel filaments, or filaments radiating out from the centre    **49**

**48b** Colony made up of cells not arranged in filaments    **50**

**H.29**

**49a** Colony large and visible to the naked eye, like cut blades of grass in the water. It is made up of masses of straight filaments, with gas vesicles appearing as dots and occasional slightly larger cells (heterocysts) at intervals along the filaments (H.29)
    *Aphanizomenon* (cyanobacterium)

**49b** Colony irregular in outline shape and comprising tangled filaments sometimes with dots and usually with relatively large heterocysts compared with the vegetative cells (H.30)    *Anabaena* (cyanobacterium)

**H.30**

**49c** Colony a sphere made up of filaments radiating from the centre. At the central end each filament has a heterocyst and the filament then tapers to a colourless hair at the distal end    *Gloeotrichia* (cyanobacterium)

**50a** Colony a flat plate with cells arranged in rows and columns. Cells may be blue green or pinkish purple (H.31)    *Merismopedia* (cyanobacterium)

**50b** Colony a hollow sphere with small spherical cells, often with black dots, arranged at the periphery
    *Coelosphaerium* (cyanobacterium)

**H.31**

**50c** Colony a hollow sphere with heart-shaped cells arranged at the periphery and linked by strands within the colony mucilage
    *Gomphosphaeria* (cyanobacterium)

**50d** Colony irregular with cells surrounded by mucilage    **51**

**51a** Colony an irregular mass of very small spherical cells, each with several small black dots (these are gas vesicles and the effect is caused by light refraction, not a black pigment)    *Microcystis* (cyanobacterium)

**51b** Colony less dense with cells more distantly spaced in an irregular mass. Cells slightly elongated. Black dots absent or faint    *Aphanothece* (cyanobacterium)

**51c** Colony like a net, with strands of mucilage containing the cells and large spaces in between. Sometimes with black iron deposits associated with the mucilage
*Cyanodictyon* (cyanobacterium)

**52a** Colony a brownish green irregular mass of spherical cells, storing oil and releasing it if damaged
*Botryococcus* (green alga)

**52b** Colony of distinctive shape and made up of geometric cells, whose walls persist on death of the cell. Oil is not stored                                                                  **53**

**H.32**

**53a** Cells forming a palisade or ribbon (could also be seen as a filament if long), with cells sometimes long and thin and wider at the centre and extremities (H.32), sometimes more regularly rectangular. This genus has been extensively revised, following work with the electron microscope in recent years and several new genera created from it (see Key E)
*Fragilaria* (diatom)

**53b** Cells long and thin, with slightly bulbous ends and in star-shaped colonies,    *Asterionella* (diatom)

**53c** Cells in zig-zag chains or star-shaped colonies but cells stocky and without bulbous ends                        **54**

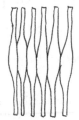

**H.33**

**54a** Partitions running part-way through the cells (H.33)
*Tabellaria* (diatom)

**54b** Partitions running all the way across the cells (H.22)
*Diatoma* (diatom)

**55a** Short filaments of a few cells loosely joined and prone to fall apart                                                       **56**

**55b** Firmly attached cells forming sometimes very long filaments                                                              **57**

**56a** Cells green, small (a few micrometres)
*Stichococcus* (green alga)

**56b** Cells brown, rectangular in side (girdle) view, circular in end (valve) view. Markings on wall in two distinct zones in end view    *Cyclotella* (diatom)

**56c** Cells brown, rectangular in side view, circular in end view. As for *Cyclotella* but markings on wall continuous from centre to edge in end view    *Stephanodiscus* (diatom)

**H.34**

**57a** Green, very thin (a few micrometres); prominent banded chloroplasts  *Gleotila* (green alga)

**57b** Brown or yellow-brown, rectangular in side view, circular in end view. Cells joined by overlapping short teeth or long spikes, usually with an obvious band across the middle of the cell in side view (H.34)  *Aulacoseira (Melosira)* (diatom)

**57c** Blue-green, blackish purple, red, grey or pale, but not green or brown  **58**

**58a** Filament helical (like a corkscrew)  **59**

**58b** Filament straight  **60**

**59a** Crosswalls prominent between cells  *Arthrospira* (cyanobacterium)

**59b** Cell crosswalls apparently absent  *Spirulina* (cyanobacterium)

**H.35**

**60a** Cells all of the same type (end cell may be slightly different shape)  **61**

**60b** Some larger, less pigmented, cells (heterocysts) inserted along a filament of smaller cells  **64**

**H.36**

**61a** Filament covered by a sheath that projects at the ends. Straight or coiled (H.35)  *Lyngbya* (cyanobacterium)

**61b** Filament not covered by an obvious sheath. Filament straight  **62**

**62a** Cells cylindrical to barrel-shaped, constricted at cross walls. End cell may be more rounded but otherwise not different from the others (H.36)  *Pseudanabaena* (also called *Limnothrix*) (cyanobacterium)

**62b** Filament not constricted at cross walls  **63**

**63a** End cell of different shape from others, filaments broad, prominently coloured blue-green or pink  *Oscillatoria* (cyanobacterium)

**63b** Very fine filaments. Gas vesicles prominent giving granular, greyish appearance  *Planktothrix* (cyanobacterium)

**64a** Filaments very straight and fine, heterocysts not much bigger than other cells  *Aphanizomenon* (cyanobacterium)

**64b** Filaments rarely dead straight; heterocysts prominent in size and colour  *Anabaena* (cyanobacterium)

## 4.3 Coping with predators

The rapid production of new generations, by rotifers and water fleas, means that changes in shape and size can be made quickly in relation to cues from the presence of predators. As the summer progresses, and young fish hatch and invertebrate predators become more abundant, it is not unusual for rotifers to acquire extra long projections, or water fleas changes in body size and shape, to frustrate detection by fish and handling by invertebrate predators. The more awkward the handling, the greater is the chance of escape. Water fleas may become smaller and more transparent, and may reduce the size of their eye to make themselves less easily seen by fish. They may also change their behaviour by developing abilities to migrate down to deeper, darker water by day or into and out of beds of plants, where they may find refuge. Fish too may change body shape or number and length of spines to evade predation from piscivores (fish-eaters), but this has to occur over several years through natural selection in the presence of their predators. The responses of the phytoplankton, through sometimes producing larger and spinier colonies in the presence of grazers and parthenogenetic zooplankton, can be much more rapid and the selection has been for the ability to change rather than the details of the change.

It is easy to grow up a mixed algal culture for the purposes of feeding zooplankters, by adding a pinch of houseplant fertiliser to a jugful of pond or rain water. To do experiments on algae though, really needs single-species cultures. Some physiologists would argue that pure (axenic) cultures, lacking not only other algae but also bacteria, are necessary, but these are very difficult to create and maintain, and one can never be entirely certain they are bacteria-free. Moreover, associated bacteria are sometimes needed for the production of vitamins that ensure normal development of the algae. Axenic cultures often have misshapen cells compared with those in the wild. The usual compromise is to accept the presence of bacteria but to work with just one species of alga.

There are various ways of isolating a single species, the most practical of which, without special facilities, is to use a fine pipette to pick up single cells or colonies under a microscope and transfer them to sterile media (best prepared by boiling then cooling some pond water or rain water to which a small amount of fertiliser has been added). The cultures should be kept in bottles or tubes, loosely stoppered with rolled cotton wool, or a screw cap. Once the culture has grown it can be sub-cultured into new tubes every so often by transferring a drop to new medium. Experiments are then possible by adding water in which water fleas have been kept and investigating whether this induces changes in shape or colony size. So far there have been relatively few such experiments and we really do not know how widespread the phenomenon is.

Likewise, zooplankton cultures can be grown by picking out individuals (egg-bearing in the case of copepods) under a stereomicroscope and placing them in pond water that is provided with a daily shot of food from a mixed algal culture kept for the purpose. An addition of about 1 ml per day from a visibly green algal culture to 100 ml of zooplankton culture should be enough. You can then test the effects of adding water from an aquarium in which fish have been kept, or from cultures of predators like the raptorial water flea, *Polyphemus*. Little is known of responses of copepods to the presence of fish in freshwaters. Establishing a culture of copepods is more difficult because they are not parthenogenetic, but one might be started from several individuals or some egg-bearing females. Experiments are also possible on growth and reproduction rates of zooplankters using different algal cultures with organisms of different sizes to test some of the general principles adduced earlier in this chapter.

**raptorial**

grasping, as in raptors (hawks, falcons, eagles) among birds

## 4.4 Geography and time

There are big differences among the plankton communities of different lakes and ponds, and these can be investigated relatively simply using net samples. Nets will be selective towards the larger species, but it is these that form the more persistent populations. The tiny species tend to increase rapidly and be rapidly grazed so that their populations are ephemeral like those of annual garden weeds. The bigger species will form peak populations at certain times of year but will generally always be found in concentrated samples. They also offer the opportunity of looking at the wide range of fungi and protozoans that colonise their bodies, sometimes just as a surface to settle on but often as parasites when they extend mycelia or other processes into the alga and eventually kill the cells.

If you have access to a group of ponds or lakes and can sample them all with a net in winter, spring, summer and autumn you can build up a picture of how the local geology and hence water chemistry influences

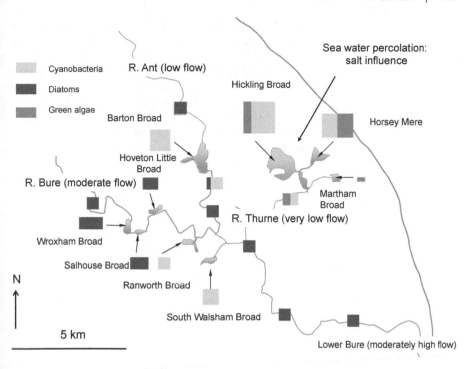

**Fig. 4.6** Pattern of phytoplankton communities in the Norfolk Broads in summer in the 1970s. The communities are mixed but only the predominant groups are shown for simplicity. The Rivers Ant and Bure received nutrients from sewage effluent and agriculture; the River Thurne received them from agriculture, and Hickling Broad and Horsey Mere also from a roost of black-headed gulls. All sites thus were nutrient polluted and when flows were low or water was retained for long enough times (in the Broads), this was associated with predominance of cyanobacteria. The River Thurne received salt water from percolation through the coast and then through pumped drainage of the land, and this was associated with a partial predominance of green algae. In the rivers, and Broads close to them, when they were rapidly flushed by the River Bure, diatoms predominated.

**hard waters**
those with high concentrations of ions, particularly calcium and bicarbonate. The origin of the term is obscure but may relate to the difficulty (hardness) of making a soapy lather with such water.

the community and also how it changes with season. You can gain a rough idea of how big the community is by using a Secchi disc (the lesser the transparency, the greater the algal community unless the water is muddy) or from the volume of the algae in the net sample if you standardise the amount of water passed through the net, and allow the sample to settle over a set period in a narrow container. In general, in soft, low calcium, low pH waters, you will find desmids and diatoms, usually with circular genera of the latter predominating. In harder waters you may find more cyanobacteria, and diatoms that are bilaterally symmetrical (long and thin cells), and in very heavily nutrient-enriched waters, many small green algae, but there are complications, High flushing rates favour diatoms over cyanobacteria (Fig. 4.6); con-tamination with farm wastes may favour euglenoids. Over a year or two you should be able to build up a picture of the algal vegetation which can be viewed in the same way that you would look at the land vegetation of an area and the influences on it. With the benefit of a microscope, the plankton can be very rewarding.

# 5 Catchments, nutrients and organic matter

There is a famous saying in the environmental world that 'All things are connected'. It is alleged to have been said in 1854 by Chief Seattle when he was being harangued, by a government agent, to sell his lands so that a railway might be pushed through to the west coast of the United States. His point was that, to his people, all features of the land and its wildlife meant something, but to the European colonists there was no such deep significance. Though there is doubt that Seattle actually said these precise words, the saying has been adopted to reflect that in the natural world, through the movements of water and the substances it carries, the dependence on them by all living organisms and their own interdependence through food webs, symbiotic, predator/prey and host/parasite relationships, the watchword is connection not separation.

Living on islands that are so populated that the land has been intensively managed for thousands of years, and where the remaining vegetation has been isolated as relict clumps and largely replaced by walled, fenced or hedged

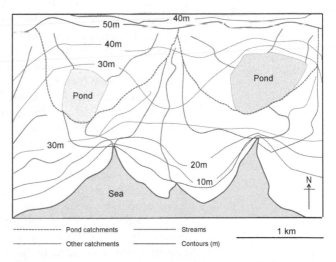

**Fig. 5.1** Catchments are the areas from which water drains to a particular point: a pond or lake outlet or the sea. The catchment areas of the two ponds here are shown in red, the catchments of streams not draining to ponds in black. The catchment boundaries are sometimes called the watersheds and the catchments themselves also watersheds in North America or river basins in Europe.

fields, the concept of separation comes more easily. So it was that the early ecologists in the UK referred to the woodland, the heath, the bog, the river and the pond ecosystems, rather than to the catchment, watershed or river basin, which is the truly natural unit, and perhaps did a grave disservice to the development of sympathetic management of a landscape that remains subtly connected, though artificial boundaries have been imposed upon it.

A catchment (Fig. 5.1) can be the area of land from which rain and snowmelt drain into a single river system, either over the land surface or through its underlying rocks and its soils, eventually to reach the ocean at a single point, albeit there may be formation of a many-channelled delta, by which the river negotiates its way through the silts and sands it has deposited following erosion of its catchment. It can also be the land area from which water drains to any single point: a pond outlet or inlet for example. All of the land surface is made up of catchments, separated by divides on the higher ground where rain moves one or other way into separate river basins. With underground waters, moving through porous rocks, the surface catchments may be blurred at the edges, but this does not undermine the concept. A lake or pond therefore is not isolated. It is like an organ in the body through which the bloodstream passes on its way to the heart from which the fluid is pumped around again. The evaporation of the ocean acts as a main pump in forcing the water back as rain and snow into the bloodstream, the freshwater system, of the land. Indeed a map of a river system looks exactly like a diagram of the body's circulation (Fig. 5.2).

Blood connects and gives and takes, and so does water. As it moves across and through the land, it dissolves substances from the soil and the rocks, and, depending on the vigour of the flow, picks up organic debris and mineral particles. All of these will have ecological effects throughout the system, whether the water moves on, or is retarded for a time in a basin such as a pond or lake, a bog or marshy wetland, the floodplain of a river or the water ponded back in the estuary by the salt tides of the ocean. The water chemistry is the medium for growth, and the deposited sediment, the basis for plant rooting in a pond. Organic matter, dissolved, or as soil, leaf or woody debris, provides much of the food energy to overwhelm or at least supplement that produced by photosynthesis in the water. In turn there is a backflow to the land, whereby energy and nutrients are returned in the bodies of emerging

**Fig. 5.2** Freshwaters look like a bloodstream and act like one in relation to the land surfaces. Delta of the River Lena, in Russia, showing large numbers of braided streams and small ponds. Photograph by NASA.

insects that feed spiders, birds and bats in the hinterland, and when land animals, like wading birds, peck invertebrates, otters catch fish, or bears and eagles scavenge the bodies of exhausted salmon on their migrations back to their spawning grounds from the ocean. Alas this latter is no longer much seen in Europe for lack of bears. There are thus three processes to be examined in understanding the linkages between a pond and its catchment: the movement inwards of dissolved inorganic nutrients, the role of organic matter washed in from the land and the reciprocal transfers back to the land, and I will consider each of these.

## 5.1 Conductivity, pH and major ions

A useful and inexpensive instrument in all this is a conductivity meter. Equally handy are some pH (acidity) indicator papers covering a range from about 3 to 11, or with greater precision, from 4–8. The meter measures the electrical conductivity and thus the total amount of charged ions in a water. These include the predominant positively charged ions (cations) $H^+$, $Na^+$, $K^+$, $Ca^{++}$, $Mg^{++}$ and the predominant negatively charged ions (anions), $Cl^-$, $SO_4^{--}$ and $HCO_3^-$. There are, of course, many other ions at much lower concentrations, but these eight, called the major ions, dominate the conductivity and also tell us a great deal about the origin of the water that has supplied the pond. Separately they are mostly expensive to analyse and impossible in the kitchen. It is however, possible to buy kits designed for field analysis, but these are expensive and sometimes not sensitive enough to measure the concentrations in ponds. They are mostly intended to assess the much greater concentrations in industrial effluents. Much can, in any case, be deduced from a conductivity and a pH measurement.

Waters that flow over and through igneous and most metamorphic rocks (the exception is marble) and sandstones tend to have a low conductivity, perhaps lower than 100 microsiemens ($\mu$S) per cm. (The concept is that of a potential current moving between two electrodes placed 1 cm apart). Such waters are predominant in the uplands of northern and western Britain and western Ireland. Rain itself has a conductivity of up to about 40 $\mu$S per cm because it picks up carbon dioxide from the atmosphere, which reacts to form bicarbonate ions ($HCO_3^-$) and hydrogen ions ($H^+$), and also acquires droplets of sea spray and dust, some of which is soluble. Poorly weatherable rocks add little to this and the solution is dominated by hydrogen (from reaction with carbon dioxide) and sodium and chloride ions (from sea spray droplets). The pH of such waters thus tends to be on the acid side of neutrality with a range from about 5.3 to 6.7 unless the rain has been unduly contaminated by acid gases when the stream or pond water derived from it could have a pH as low as 3. Remember that pH is a logarithmic unit and a difference of 1 in pH means a tenfold change in hydrogen ion concentration. There is a range of over one million in the hydrogen ion concentrations of natural waters, a billion if polluted waters are concerned.

In contrast, ponds supplied from catchments with shales, mudstones, other fine-grained sedimentary rocks and glacial drift (and indeed ponds in cultivated gardens unless they are raised up and fed only by rainwater) will have conductivities between 100 and 300 $\mu$S per cm, and a pH straddling neutrality, from around 6.5 to 7.5. In limestone and chalky areas, either upland or lowland, or in regions like the glacial plain of Cheshire, Staffordshire and Shropshire in north-west England, where the glacial deposits may contain a lot of limestone fragments, the waters will be dominated by calcium and bicarbonate ions, have conductivities up to 600 $\mu$S per cm and pH values normally around 8, but sometimes up to 10 or even 11. Close to the coast, if there is infiltration by seawater, ponds may have conductivities in the thousands and pH values around 8 because of a dominance by seawater. Blown sand contains a lot of shell fragments that steadily dissolve in pond waters and it is instructive to investigate ponds in sand dune systems at different distances from the sea. Their conductivities in theory fall rapidly away from the coast.

A simple knowledge of the local geology, even if you do not have a conductivity meter, will tell you a great deal about the chemistry of your pond (Fig. 1.4) and of the plants and animals you are likely to find in it. Snails, and the larger crustaceans like crayfish and pond shrimps, require a great deal of calcium for their shells and exoskeletons, and are scarce in low conductivity, acidic waters. Many plants can use bicarbonate as a source of carbon dioxide for photosynthesis, whilst others can only use the carbon dioxide molecule. The latter include the shoreweed (*Littorella*), water lobelia (*Lobelia*) and the isoetids (*Isoetes*) (Fig. 5.3), all of them low-growing plants in acid, soft waters, low in bicarbonate

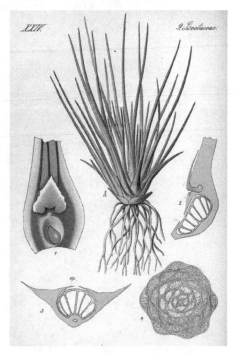

**Fig. 5.3** *Isoetes* (quillwort) is typical of a group of plants that grows in very low carbonate waters and uses carbon dioxide from the sediment, which it absorbs through its abundant roots. Other genera that do this include *Lobelia* and *Littorella*. Picture taken from Otto William Thome's Flora of Germany, Austria and Switzerland, 1864.

and limited by the relatively low concentrations of free carbon dioxide that can dissolve in water in equilibrium with the air. They are outcompeted in neutral and alkaline waters by the more vigorous pondweeds and elodeids that can use the much more abundant supply of bicarbonate. A study of the nature and abundance of water plants or snails in relation to conductivity and pH and local geology in a suitably large area can be rewarding as a way of revealing other influences that cause deviation from this simple relationship. Desmids (see the keys in previous chapters) other than some species of *Cosmarium*, *Closterium* and *Staurastrum*, also cannot use bicarbonate and a study of the bottom algae in an area that ranges from acid to alkaline waters will give you great pleasure. Desmids are very beautiful (Fig. 5.4).

Geological maps and bottled water can also tell you a great deal. The maps need interpretation for they show strata by their geological age rather than by their nature, but

**Fig. 5.4** Desmids come in an intricate variety of shapes but basically have cells in two halves separated by a groove or isthmus that may be deep or barely visible. Painting by David Williamson.

igneous rocks, chalks and limestones are readily picked out and the handbooks that accompany the maps will give more details. Bottled waters often have the location of their origin, and some chemical analyses, printed on their labels and can give clues to local water chemistry, but beware those produced by large soft drinks manufacturers, who blend water from several sources (but usually omit telling you the chemistry anyway). The British Geological Survey (Cribb & Cribb, 1998) in an uncharacteristic moment of whimsicality once produced a booklet relating geology to the tastes of single malt Scotch whiskies. There is a link between the rather phenolic tastes of those that are produced in the peaty acidic waters, with their bog pools, of the Hebridean Western Isles, compared with the smoother, less raw tastes of those made with high conductivity waters in the east of Scotland. The BGS booklet was not so convincing about the details of the intermediate varieties and further research is merited.

**phenolic**
having the characteristics of phenol (carbolic acid) or similar compounds

## 5.2 Nitrogen and phosphorus, key nutrients

A second aspect of the nature of the catchment concerns two essential nutrients, the compounds of nitrogen and phosphorus. The major ions discussed above are also essential to living organisms but, with the occasional exception of carbon when stocks of carbon dioxide have been temporarily used up by photosynthesis on sunny days and take time to be replenished from the atmosphere, the supplies of all of them, even in the lowest conductivity waters of the mountains, tend to be much greater than the needs of the organisms. That is often not so for nitrogen and phosphorus compounds. You can investigate this if you can obtain some sodium or potassium hydrogen phosphate, some ammonium chloride or ammonium sulphate or some sodium or calcium nitrate. There was a time when you could buy these from the local pharmacists, but a combination of central production of medicines and an over-zealous attitude to safety have got in the way of this. You can but ask! Close inspection of the cartons of fertiliser at a garden centre, or contact with a friendly farmer, may provide an alternative. Many fertilisers and houseplant feeds contain both nitrogen and phosphorus compounds and are less useful, though they can be used in a general way.

You need to prepare solutions of these compounds so that when you add a small quantity (say 1% of the volume of water) to a sample of stream or pond water, you will be adding about 250 μg of phosphorus per litre or 1 mg of nitrogen per litre. This needs a little calculation with atomic weights, a kitchen scales and a measuring jug, but if all else fails add a teaspoonful to a bucket of water. Old jam jars are useful experimental pots and for each water you test, you need at least four: one with just pond water to act as a control, one to which you add phosphorus, one nitrogen and one both nitrogen and phosphorus. In case of oddities (the jar might be toxic because someone used it once with paint remover, or you forgot to add the nutrient, despite being totally convinced that you did), you should have three replicates of each treatment, a total of twelve jars. Try a range of ponds from different situations. Leave the experiments in a well-lit place (but not direct sunlight when they may heat up too much) and after a few days you may notice that the water in some jars has gone greener than in others. It is useful to have a scoring system, prepared perhaps by using watercolour paints to create a colour chart of grass-green ranging from barely visible green to pale green. It is also possible to filter the water through filter paper, which can be dried (in the dark) and used for colour comparison. The green in the water will usually be due to one of the small green algae (chlorophytes) that are very common in ponds and resilient enough to survive life in jam jars as well.

Alternatively, you can use common duckweed (*Lemna minor*) (Fig. 5.5) as a test organism. It too is very amenable. Duckweed produces flattened leaves (joints) that float on the surface. It also produces a single root extending down from each joint. One small plant with maybe three or four

**Fig. 5.5** Common duckweed (indeed all duckweed species) are very useful experimental organisms. Photograph by Barbirossa.

joints is sufficient to start the experiment, but all the plants should be a healthy green (not whitish or yellow) and should come from the same source. There are three things that you can measure that may reflect the nutrient status of the water: the total number of joints produced, and their rate of increase as the plants grow and bud off new ones, the size of the joints, measured in a standard way and the length of the root. If you have enough replicates you can also use simple statistics such as one-way analysis of variance to test whether there have been significant differences among treatments, or regression analysis to determine the rate of change in numbers or sizes of joints or root length under different conditions. There are various books that discuss statistical tests, including one in this series (Wheater & Cook, 2003) or you may have a program on your computer that will do the job for you. But often the results will be obvious without the need for statistics.

Your experiments will generally show that when you add phosphorus, or both nitrogen and phosphorus, there will be an increase in greenness, or more duckweed produced. Sometimes nitrogen alone will stimulate growth (Fig. 5.6). What it means is that phosphorus or nitrogen or both were

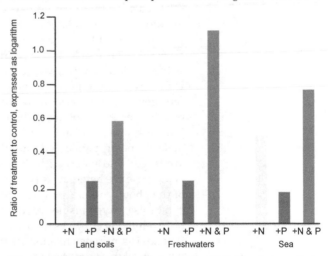

**Fig. 5.6** Experiments can be carried out in which phosphate or nitrate (or ammonium) salts or both are added to soils or waters and growth of plants or algae followed over the next few days. The amount of growth in the treatment is divided by that in the control (to which no nutrients are added) and expressed as a ratio. For convenience this ratio is expressed as a logarithm. Thus a value of 1.0 on the vertical axis in this diagram means a tenfold increase in growth. Results are averaged here for several hundred different experiments on soils, freshwaters and seawater across the world. Based on Elser *et al.* (2007).

limiting to growth in the original water. Both elements are relatively scarce in the biosphere in available form. Phosphorus is absolutely scarce because of the accident of how the original gas and dust cloud from which the Earth formed was made up, but also because its compounds are relatively insoluble and it is readily bound to clay particles in soils. Land plants use a lot of energy, and often must enter into associations with fungi, called mycorrhizae, to obtain enough phosphorus. Consequently, natural vegetation tends to conserve phosphorus in its biomass, leaving rather little at risk of being lost to leaching by rain for delivery to streams and ponds.

Nitrogen is very abundant as nitrogen gas but only available in this form to nitrogen fixers, all of which are bacteria. They convert nitrogen gas to a series of compounds and finally to amino groups that characterise amino acids and the proteins they form. Decomposition of bacterial proteins releases ammonium ions ($NH_4^+$), and other bacteria oxidise ammonium to nitrate ($NO_3^-$), from both of which other, non-fixing organisms can obtain their supplies. Nitrogen-fixing bacteria require low-oxygen conditions to function. They occur in soil, surrounded by clusters of other bacteria that mop up oxygen from their immediate vicinity, or inside nodules on the roots of a limited number of plants, notably the legumes, alders and buckthorns. Sometimes haemoglobin is produced in the nodules, also with the role of reducing oxygen tensions around the fixers. In freshwaters, one group of cyanobacteria fixes most of the nitrogen. Cyanobacteria can occur as single, very tiny cells or as filaments or colonies where all the cells are similar. But one group has filaments that have two sorts of cells, the majority being pigmented and photosynthetic, but some are larger, and paler, with the contrast of a buckle on a belt. These are heterocysts (Fig. 5.7) and lack certain parts of the photosynthetic pathway, so that they do not photo-synthesise, oxygen is not produced within them, and the nitrogen-fixing enzymes can function. Indirect measures of potential for nitrogen fixation are the number of heterocysts per unit volume or percentage of them of total cyanobacteria cells or of total algae in the sample. This can be tested by combining such measures with the experiments on nutrient limitation described above.

Nitrogen compounds are ephemeral because when proteins are decomposed to ammonium and then nitrate, they are not only available to plants and algae but also to even more bacteria that can oxidise ammonia to nitrogen

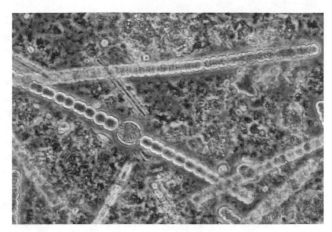

**Fig. 5.7** Plankton settled from a very shallow lake with much resuspended sediment. The prominent long filaments are of a cyanobacterium, *Anabaena* species, which has two sorts of cells. The larger, apparently empty, cells are heterocysts in which nitrogen fixation can be carried out because a low oxygen environment is maintained within them. The sediment shows that the filament is surrounded by a layer of mucilage. Such capsules are common in prokaryotes.

gas ($N_2$) thus releasing energy for their growth, or which can use nitrate in deoxygenated, waterlogged soils, or in water itself, to oxidise organic matter to release its energy. All these processes result in the combined nitrogen eventually, and sometimes very rapidly, being converted back to nitrogen gas and are the reasons why nitrogen is often found to be limiting despite the huge reserve in the atmosphere.

## 5.3 Eutrophication

In the remote past, before the landscape was converted from natural vegetation to agriculture and pasturage, it is likely that both nitrogen and phosphorus simultaneously limited growth of algae and plants in freshwaters, a situation that still pertains in areas of the ocean. But from the start of the twentieth century, development of the Haber-Bosch process produced ammonia industrially for subsequent conversion to nitrate fertiliser. This process is now releasing about as much combined nitrogen to the biosphere as nitrogen fixed by natural processes. In consequence there is a surfeit of nitrogen in agricultural areas, and because both ammonium and nitrate are very soluble in water, this places greater emphasis on the scarcity of phosphorus in freshwaters. But phosphorus concentrations have also risen as a result of human activities, because of use of super-phosphate fertiliser and because human and farm animal sewage is rich in phosphorus and nitrogen. When human populations were lower and most sewage was redistributed to the land, much of this phosphorus was refixed in the soils. Distribution to the land in densely populated areas is no longer acceptable and the sewage is treated to remove organic matter, leaving an inorganic solution of nitrogen and phosphorus compounds and a large array of other substances that pass through the works. As a result of both fertiliser use and large human populations, there is now a severe problem of nutrient pollution, or eutrophication, worldwide and concentrations in rivers, ponds and lakes are, on average, about ten times as high as they would be were the land undisturbed and lightly populated. Nonetheless, experiments such as those described above will still show stimulation of growth by these nutrients in many waters. It is worth trying to emulate natural eutrophicated conditions by doing your experiments in low light. Below the surface, light is scarce in eutrophicated ponds because of absorption by algae and the large amounts of dissolved organic matter that reach such waters. Do you get different results from the same addition of nutrients in normal daylight compared with shaded and deeply shaded conditions?

Garden ponds, and ponds on farms where stock is kept, are usually highly eutrophicated. Fertiliser is liberally used in gardens, be it from an expensive bag or the local stables or chicken farm. Mixed and stock farms can hardly avoid a liberal dressing of manure on their fields and although manure piles from cowsheds and stables are supposed to be covered from rain, it is rare for there not to be a dribble of drainage that does not find its way into a ditch or culvert and down to some of the ponds on the farm. If the garden pond is built up above ground level and lined, it will not be so nutrient enriched for its catchment is limited to its edge and its nutrient supply will come from rain and from topped up tap water in dry periods. Rain however is now relatively rich in nitrate from oxidation in the atmosphere of ammonia volatilised from the manure of intensive animal farms, and from the burning of petrol and diesel in vehicle engines, which release a variety of nitrogen oxides that oxidise in the atmosphere to nitric acid. Tap water is also usually rich in nutrients. Ultimately it comes from ground or river waters that have received nutrients from the land or sewage effluent and although recent legislation has been passed to require removal of nitrogen and phosphorus from the effluents of large treatment works at least, there are still large residual concentrations. The nutrient status of your pond will depend partly on the local geology, for soils derived from sedimentary rocks are richer in phosphorus than those from igneous rocks, but the contrast is not nearly so great as that between water derived from natural or semi-natural vegetation and that derived from farmed soils. It is land use, whether in the wider countryside or in the immediate garden that determines the productivity of a pond. You can test this using different stream waters and the sorts of experiments described above.

## 5.4 Supplies of organic matter

The catchment provides more than dissolved nutrients; it also provides organic matter and that organic matter, either dissolved in the water, present as fine particles eroded from soils, or as leaves, bud scales and twigs, is an important energy source for both flowing and standing waters. In streams, large branches and collapsed tree trunks are important in creating a structure that retains smaller debris. This supply of organic matter (Fig. 5.8) is most evident in headwater streams in areas still retaining woodland or forest. Overhanging vegetation darkens the small streams and little photosynthesis is possible. Most of the tree debris enters in autumn and winter when nutrients have been translocated back from the leaves to the branches and so the leaf debris is poor in soluble nutrients and indigestible to animals. There is, however, a group of fungi in the water, called the Hyphomycetes that specialises, with spores shaped like anchors or hooks, in colonising such leaves. Like the mycorrhizae in forest soils, the hyphomy-

**Fig. 5.8** Large amounts of tree and leaf debris can be normal parts of many ponds, providing structure and sources of organic matter and energy to the invertebrate community.

**mycelium (plural mycelia)** the vegetative (non-reproductive) part of a fungus

cetes are efficient at gleaning nutrients from solution and as their mycelia permeate the leaves, they convert the indigestible cellulose and lignin to much more palatable protein. It is a little like the improvement of a commercial bread roll by a slab of hamburger meat. Invertebrates in the stream, classed as shredders and including various fly larvae and crustaceans, and in the tropics prawns and crabs, then tear the leaves apart to obtain the mycelia. As they do so they break up the leaves and produce copious faeces that become colonised by bacteria and add to the food of filter collectors, notably the blackfly larvae and various caseless caddisfly larvae, or, where the fine material settles in quiet places of the stream, deposit feeders like oligochaetes, chironomid larvae, and bivalve molluscs. Dissolved organic matter entering from the catchment feeds a film of bacteria and protozoans on the stones and tree debris, and together with any algae that can manage to grow, this film supports scrapers like snails, mayfly and stonefly larvae. Invertebrate predators (leeches, dragonfly and damselfly nymphs, diving beetles and water bugs) and fish feed on these guilds of invertebrates.

Trees often overhang ponds and leaf debris from trees and grasses is also an important source of food. Shredders and ultimately deposit feeders feed on it, but passive filter collectors are scarce, in the absence of a current to bring the material into their filtering limbs or nets. Tree debris will also support a scraper community and sometimes the leaf debris will dominate the functioning of a pond, as in ponds set in the middle of woodlands. The amount of debris, coupled with the shortage of oxygen, in the absence of a vigorous flow, may create anaerobic conditions over much of the bottom and provides for a distinctive community, particularly of fly larvae and sometimes a visible colouring of anaerobic, photosynthetic purple sulphur bacteria (Fig. 5.9).

## 5.5 Experiments with food supplies

It is an interesting comparison to sample a set of small streams and ponds in the same area and to analyse their invertebrate communities according to the feeding guilds present and their relative abundance. Table 3.1 suggests the most common guilds for the major groups, but greater sophistication can be had by identifying to family and then using Dobson *et al.* (2012), more detailed keys and reference books to discover more subtleties. Some groups, the mayflies for example, are represented by several guilds, whereas oligochaetes are entirely deposit feeders. A chi-squared test can be used to determine whether there are significant

**Fig. 5.9** Anaerobic conditions, in forest ponds with plentiful leaf supplies, can support large growths of purple sulphur bacteria. They are photosynthetic, using hydrogen sulphide instead of water in the process. Photograph by Nick Sanderson.

differences in the proportions of different guilds in different places. Some measure of local tree coverage allows relationships to be determined between supply of leaf litter and guild structure. Simple traps, sheets of polythene for example, can be laid out to catch leaf litter over a 24-hour period, and quantify the amount coming in. Measurements should be made throughout the year, though the bulk will enter in autumn. Many leaves fall more distantly and are blown in. Low fences of chicken wire can be used to catch these and measure the amounts.

There are also possibilities for detailed observations and experiments. Leaf litter can be washed then placed in mesh bags and left for some weeks in a pond (remember to attach a string to be able to recover the bags). Fruit cage netting will allow access to all invertebrates, different sorts of net curtain or bridal veil material provide different mesh sizes that will exclude large species but not small ones. Very dense material will exclude all but microorganisms. Many windows in Wales and Scotland seem to have net curtains, so I recommend a Celtic drapers for obtaining the greatest diversity of curtain material. Cotton material does rot in time, so polyester is best, but it is also possible to buy (though expensively), nylon netting of precisely defined mesh size. Armed with 10 cm x 10 cm bags, sewn or stapled together, you have a useful tool for investigating the process of litter decomposition. Are invertebrates necessary or will microorganisms do the job just as fast? Does size of animal matter? Are different leaf species shredded at different rates?

**Fig. 5.10** The water hoglouse (*Asellus aquaticus*) is a common leaf shredder in ponds. It will also graze on filamentous algae. Photograph by Jones.

What is the half-life (the time it takes for half of the initial material to disappear) for different leaf species? Most leaves fall and accumulate in the water in autumn. Do they begin to decompose then or does the process have to wait until spring or summer? Does decomposition proceed faster with the same biomass of two or more shredder species (Fig. 5.10) compared with just one? Many people remove leaves that fall into their garden ponds. What effect does this have? Smallish garden ponds can be divided into two using polythene sheeting held down by stones and stretched on wire at the surface, allowing you to do (in a small way) whole-lake experiments. Try removing leaves in autumn from one half and leaving the other as a control. What happens to the invertebrate community and the water chemistry?

Finally, do not underestimate the importance of leaf accumulations in ponds. They are part of the sediment, either intact, or as fine material left after the shredders and deposit feeders can no longer use them and the remains become buried and anaerobic. You can see something of this importance if you place sediment in a glass jar, allow it to settle for a few hours and watch its surface. The surface in winter will be a light, often rusty brown colour if the water is well oxygenated, but if you repeat the observations under summer temperatures, you may find that this oxidised surface layer is thinner or disappears, and the sediment is dark brown or black throughout. The light brownness is due to oxidised iron compounds and these are maintained provided enough oxygen can diffuse into the sediment to compensate for the intense respiration of the bacteria. In summer, with higher temperatures and lower concentrations of oxygen in the water, even at saturation, and more intense bacterial activity, not enough oxygen may be able to diffuse in to maintain the surface light-brown layer. Reduced iron compounds predominate, sometimes reacting with sulphide produced by the bacterial decomposition of sulphate ions to give black iron pyrites (iron sulphide). Oxidised iron ties up phosphate very efficiently. Reduced iron does not, and in summer, phosphate may be released in quantity from stores in the sediment. This has major implications for the nutrient supply in the water. Phosphorus may be limiting in spring but not in summer as a result.

Burrowing invertebrates stir up the surface sediment layers and mix them with overlying water (Fig. 5.11). If you experiment with sediment in jars to which you have added animals and sediment whose animals have been removed by sieving or killed by freezing, does this make any difference

**Fig. 5.11** Oligochaete worms feeding in a deposit of fine organic matter.

to the persistence of the oxidised iron layer? Numbers of invertebrates are often reduced by fish predation (see later chapters). Perhaps fish can influence summer phosphorus concentrations through these mechanisms.

Below the surface layers of sediment, organic matter accumulates in the naturally deoxygenated conditions. Very little further decomposition then occurs, just a little by anaerobic bacteria that use nitrate or sulphate to oxidise organic matter until supplies have been completely converted to nitrogen or sulphide. Other bacteria convert some organic matter, like carbohydrates, to methane. Thereafter, the preservation of organic matter is permanent unless the sediment is reoxidised, for example by complete drying out in temporary ponds. Simply by being there, the organic sediment makes a contribution to maintenance of equable conditions on Earth but that is a matter for the final chapter.

Solid organic matter is very important in streams, perhaps less important in ponds, except in those surrounded by woodland, where it may dominate activities in the pond. Dissolved organic matter in contrast is probably rather more important in ponds than in streams. There are new techniques available involving the measurement of stable isotopes of carbon that give us information on where the carbon in different organisms came from and these are discussed in Chapter 7. The general pattern that has been obtained from small lakes and ponds is that sediment is largely derived from washed-in land and fringing reedswamp material, and that deposit feeders rely particularly on this. The zooplankters sometimes take in very large amounts of washed-in material but also take algae produced in the water, whilst bacteria in the water are usually feeding largely on washed-in dissolved organic matter. The amounts of this are much greater than the soluble organic matter released as secretions or on decomposition of the pond organisms themselves. Even in large lakes, like Loch Ness, there is still a dependence on incoming organic matter, but it is likely that in really large ones, like the African Great Lakes and in the centres of the oceans, far from the land, functioning has had to become more self-contained.

## 5.6 Net heterotrophy

Pond waters are usually oversaturated with carbon dioxide. This does not mean that the concentrations are huge. They are not because the concentration is determined by equilibrium with the air and air is not especially rich in carbon dioxide (about 400 parts per million at present, and rising) but the concentrations are often larger than they would be at equilibrium. This is because of the intense respiration of organic matter entering from the catchment. It means that ponds are often 'net heterotrophic'. More respiration goes on than can be accounted for by the photosynthetic production of algae and plants under the water surface, with the deficit being made up from what is washed in. The difference between the gross photosynthesis and the respiration of the whole community (plants, algae, microorganisms and animals) is called the net ecosystem production. If a pond is heterotrophic, the value is negative; if photosynthesis is accounting for all of the respiration, or potentially more, so that there is surplus production, the pond is autotrophic. You can determine the status of your garden pond if you have an oxygen meter. Measure the oxygen concentration (as mg per L) every hour for a complete 24-hour day/night cycle. Then calculate the rate of change in concentration per hour. It will be positive for most of the day as photosynthesis releases oxygen to a greater extent than respiration uses it up, but there will be a decline during the night because only respiration is then occurring. Plot a graph of change in concentration per hour against time and you should see a curve (Fig. 5.12) that looks like a wave, rising by day, falling by night and possibly flattening by dawn. Follow the instructions given in the figure caption and you should be able to calculate the balance of photosynthesis and respiration in the system. By repeating this several times per year, it is possible to deduce a great deal about the metabolism of the system. The estimate is only approximate because exchanges of oxygen with the atmosphere are going on but

**heterotrophy**
the condition of being heterotrophic (obtaining food from other organisms, not by photosynthesis)

**gross photosynthesis**
total photosynthesis that is taking place, but because plants and algae have maintenance needs represented by their respiration, new production is much less and is called the net photosynthesis

**autotrophic**
obtaining food from inorganic carbon, usually by photosynthesis

the correction for these is usually small unless the weather is very windy. If you choose calmer conditions, the correction can be ignored.

This chapter has illustrated how the catchment has a major influence on what goes on in a pond and its production. There is great variety in different parts of the landscape, constant change with weather and time of day. There is also change throughout the life of a pond, a topic that will be examined in the next chapter.

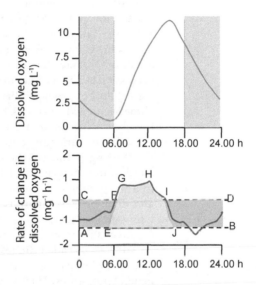

**Fig. 5.12** Determination of net ecosystem production. First, the oxygen concentration in the pond is measured every hour over a 24-hour period (upper graph). It will decline in darkness (grey) but increase in the light until the light starts to fade as night falls. Then the rate of change (increase or decrease) in oxygen concentration is determined (lower graph). The mean rate of change during the dark hours is determined (line AB) and the area coloured in pink (ABDC) is taken as the overall respiration rate of the system over 24 h. This will include the respiration of the microorganisms, animals, algae and submerged plants. The rate of change curve during the daylight hours (FGHI) is extended down to E and J on line AB and the area subtended (coloured green) is taken as the gross photosynthesis of the algae and submerged plants. The ratio of community respiration to gross photosynthesis is then ABDC divided by EFGHIJ. Alternatively the net ecosystem production is calculated as EFGHIJ minus ABDC. In this instance the respiration is clearly greater than the gross photosynthesis and the system is using organic matter imported from the surrounding land (as dissolved organic matter, eroded soil particles, or fallen leaves) to function.

# 6 The ecological development of ponds and lakes

One of the best known concepts in ecology is that of succession: the way that bare rock, mud or sand become colonised first with pioneer species that alter the initially challenging habitat and make it, by their creation of a structure, and amelioration of the soil, into one more suitable for a wider range of more fussy species. Eventually what was called a climax community ends the succession until some vigorous event: a fire, flood or windstorm, or just a natural climate change, destroys the climax and a new succession begins. Originally people believed that there was a great deal of predictability in these processes and in the climax achieved. On rocks, first came lichens, then mosses, then herbs and finally trees. We now know that there is a great deal of variability and many possible pathways from the starting point, but the idea of pioneering, generalist species, devoting most of their energy to reproduction in a risky habitat, of increase in structure and diversity, and finally of a community that has enough structure to maintain relatively stable conditions, in which specialist organisms flourish, has survived. The latter communities are still loosely called climaxes but there is no single climax, just a very large range of possible communities with some general properties. Their members have more complex life cycles, greater specialisation and many symbiotic relationships compared with the pioneer phase. Successions were also called seres, and the hydrosere, in which water was involved, was a favourite example of early ecologists. The open water was colonised first by algae and submerged plants, with reeds at the edges. Gradually the reeds built up peat soils, from their only partly decomposed remains, and encroached towards the middle, being succeeded to landward by marsh and fen plants, and then trees like willows and alder, as the soil level was raised and conditions became drier. Eventually the reeds were supposed to reach the middle and the alderwood and willow thickets to dry out and become forest, so that the once open water eventually succeeded to dry woodland. In wetter regions there was the intervention of an acid bog stage, which built up on the fen mat or in the alderwood as the peat level grew higher than the groundwater level, and became supplied mostly by low-conductivity rainwater.

There is, of course, much truth in this account and there are examples where this orderly story pans out, but there is much variation. Wet edges usually stay wet and dry woodland does not appear. The progress of the reeds is stopped by too great a water depth, or grazing by water birds on their shoots, or climate changes, over the several hundreds or thousands of years in which the succession takes place, to dryer or wetter conditions, interrupting an orderly sequence with floods and droughts. Where ponds are concerned, vegetation may be cleared to maintain open water, or intensive farming may result in so much siltation that the pond fills in quickly to a damp grassland hollow. But the concept of succession in the hydrosere remains interesting, and good examples can be seen, for example, in the filling-in of basins on the English north-west midlands plain, where there is a range of deeper meres that still have open water after ten thousand years, and shallower mosses, at various stages of the hydrosere sequence, including some with extensive alderwoods, others with wide reedswamps and many that have become completely covered with bog vegetation.

**moss (plural mosses)** local name for a bog or peat bog

There are similarities between the colonisation of water and the colonisation of bare land, but also differences. There is a much greater influence of the surrounding catchment in the hydrosere and the water itself may change chemically with circumstances but not in the predictable way that a soil in a land succession may be expected to develop. Nonetheless it is interesting to explore the theory of succession, well articulated in an article by Eugene Odum in 1961 and to combine it with other major theoretical areas of ecology. Many of the key concepts of ecology were developed from work in small lakes and ponds. There was the concept of movement of energy along food chains and the importance of detritus, articulated in a famous 1948 paper by Raymond Lindeman, based on his work in the very small Cedar Bog Lake, in Minnesota; that of the niche, developed by George Evelyn Hutchinson (1959) from observations on the coexistence of a variety of water bugs in a small pool near a shrine in Sicily and published in a widely celebrated paper called *Homage to Santa Rosalia*; and the ideas of Stephen Forbes in 1887 about lakes as microcosms, islands in a sea of land that stimulated the development of island biogeography theory in 1967 by Edward Wilson and Robert Macarthur.

## 6.1 Experiments with succession

Odum's predictions for the changes that occur in successions are shown in Table 6.1. Some of them are testable using plastic buckets in a garden, or jam jars in the bathroom window. Bathrooms are very useful in that they have frosted glass that protects small jam-jar lakes against too much hot sunlight. Buckets and jam jars are also cheap and the value of experiments is greatly increased if there is as much replication as possible.

If you sample a set of ponds that are close together you may find that their communities differ greatly but you can never be certain if this is due to the random chance of whatever species reached them first, and thence outcompeted later arrivals, or to environmental differences among them. There is a large diversity of possible colonisers (some thousands of freshwater invertebrates and freshwater algae and some hundreds of plants). The question can be tackled using experimental ponds in buckets, but they have to be set up carefully to give valid results.

The buckets have to be uniform but they have to mimic real ponds. Animals colonise ponds in three ways: by crawling in from another pond through waterlogged ground, or drier ground during and just after rainstorms, when vegetation is wet, by flying in as adult insects (or as breeding insects by laying eggs in the water), or in the guts or on the feet or feathers of water birds. They can be blown in as dust containing resting spores or eggs, but mainly in dry regions. If we imagine the landscape as the ice retreated at the end of the last glaciation and ponds and lake basins were exposed and filled with rain and melt water, we can design suitable bucket ponds. They should be of inert material, so of polyethylene not metal (or you could use ceramic plant pots if they are glazed on at least one side and you block the holes in the bottom). They should be buried so that their rims are flush with the soil surface and the surroundings are the same for each replicate. They should have a substratum of washed sand (don't use garden soil-it is very difficult to prepare a uniform soil sample and the soil may contain too much nutrient and pesticides); and they should be filled to the top with rainwater that has been collected in a plastic tub and preferably from roofs with plastic gutters. To be statistically correct you should not only put several replicates (minimally three but five is a good number) in one place in your garden, but a second and preferably third set in one or two other locations. This is to account for the fact that any single location has specific

**Table 6.1** Changes that occur as ecosystems develop from the pioneer to the mature state.

| Characteristic | Pioneer | Developing | Mature |
|---|---|---|---|
| Organism size | Very small | Wider range | Very wide range |
| Biomass | Very low | High | Very high |
| Structure | Very low | Increasing | Very high |
| Production to Biomass ratio | High | Falling | Low |
| Production to Respiration ratio | Low then high | High | Approaches one |
| Net production (gross production minus respiration) | High | Falling | Zero |
| Food chains | Simple, grazing | More complex, grazing | Web like, grazing and decomposer |
| Available nutrients | High | Falling | Very low |
| Recycling and nutrients bound in biomass | Low | Increasing | Very high |
| Nutrients leachable from the system | High | Falling | Very low |
| Biodiversity | Low but increasing | Variable | Very high |
| Niche specialisation | Low | Increasing | High |
| Life histories | Simple, geared to production of many offspring | Simple to more complex | Complex, geared to production of fewer offspring with high chance of survival |
| Mutual symbioses | Few | More | Many |
| Summary characteristics | Establishment, Survival | Production, Growth, Quantity | Protection, Stability, Quality |

characteristics (shade, shelter, leaf blow from trees) that might influence the processes of colonisation.

With a good set-up, the scope for experimentation is wide. You can see whether season has an effect, whether putting in some organic debris has particular attractions, whether peat (soak it first or it will float) or gravel have different influences and whether nutrient state (add a little fertiliser) affects the establishment of the community. If the powers-that-be object to your digging holes in the lawn, explain that lawns are basically very uninteresting parts of gardens and that holes can be repaired. Or if opposition is fervent, and you have a microscope, use glass jam jars, and concentrate on the colonisation by algae and protozoa alone. Larger invertebrate animals will generally not be attracted to such small pots but algae and protozoa cannot make choices. They drop in from the air or come in rain.

Such experiments should last for several months. Nothing may happen for some time, or there may be immediate colonisation, but such variations give insights into the process. You can sample animals using a small

**Fig. 6.1** The author thrusting a plastic drainpipe into the sediments of a shallow lake to obtain a core. The pipe had previously been cut longitudinally and then re-taped together with strong waterproof tape that could be cut on land to retrieve the sediment. An intact tube can be used but then a plunger is needed to push the sediment out of the core. Modern attitudes to health and safety would now mandate the wearing of a life-jacket.

home-made hand net or a coffee sieve, or small nets made for aquarium enthusiasts, used in a predetermined standard way, but you must replace the animals after examination and counting. Algae pose more problems but if you take a pipette and suck up a small volume of water from the bottom, where they will settle in small containers, you should have something to look at. If you take a sample from the body of water you will probably find little. The bottom is where the action is in very small ponds.

As your systems develop, the prediction is that things will change in the ways suggested in Table 6.1 and you should be able to test at least some of them. The system should become more structured (leaves may fall in for example) with time and filamentous algae, even plants, will also provide structure. You can make measurements on organism size and if you identify the animals and use the handbooks you will be able to measure changes in feeding guilds. Diversity is easy to measure and identification and reference books will reveal something of changes in life history and niche specialisation. When, after several months, your pond systems have established, they will be usable for testing other aspects of ecological theory, but first, what do we already know about development of lake and pond systems?

## 6.2 Palaeolimnology and the ecological history of natural ponds and lakes

There is a whole area of freshwater science, called palae-olimnology, which specialises in discovering the ecological histories of lakes. The raw material is an undisturbed core of sediment, usually taken with a sophisticated corer especially if the water is deeper than wadeable. In ponds, it is possible to push a plastic pipe into the sediment and sometimes possible to pull it out without all the sediment being lost. I have done this (from a boat, using a drainpipe about 10 cm diameter, sliced lengthwise and then retaped together) in the Norfolk Broads, where a plug of the underlying peat was sufficient to hold in the core (Fig. 6.1). Success is improved if the pipe is completely full of water and sediment, and can be capped in some way before being pulled out. Typically several metres of sediment can be sampled by a specialist corer from a glacial lake, covering the past ten or more thousand years. From the Broads I obtained about a metre of sediment spanning their history from their origin in the thirteenth century. From an old pond, with some trouble and by getting wet and muddy, you might get half a metre, but that is enough to demonstrate some principles and give some clues. Cores can be extruded from the pipe using a plunger to push the sediment out. As it emerges you can slice it into sections of 0.5 or 1 cm with a palette knife, and store it in small plastic bags in a refrigerator, or freezer if you cannot examine the sediment within a few days.

A huge range of measurements can be made on sediment, in a fully equipped laboratory, to reconstruct lake history. First the nature of the sediment (organic content, grain size, carbonate and water content) can be determined, then the chemistry, from the main ions and nutrients to a huge range of organic compounds derived from plants and algae. Examination of small quantities in water with a stereomicroscope will reveal seeds, leaves or fragments of epidermis, and the star-shaped sclereid cells of water lilies; there may be parts of the exoskeletons of crustaceans, sponge spicules, head capsules of flies, wings of beetles and shells of snails (Fig. 6.2). With specialist preparation and a compound microscope, pollen grains, the cysts of yellow-green algae, sometimes remains of cyanobacteria and usually the siliceous walls of diatoms, will often prove abundant.

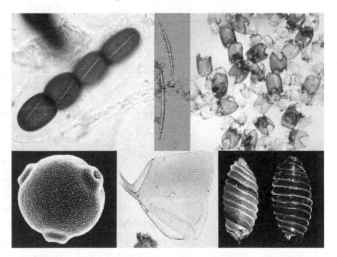

**Fig. 6.2** Some of the sub-fossils (in addition to diatoms, which are shown in Fig. 6.3) that can be found in sediments and used to reconstruct ecological history. Top left are spores of a dung fungus, *Sporormiella*. Top centre are siliceous freshwater sponge spicules and top right are head capsules of chironomid fly larvae. Bottom left is a pollen grain of birch. Bottom centre is the remnant of the carapace of a zooplankter, *Bosmina* and right, oospores of *Chara*. Photographs by Yarrow Axford, Anna de Sellas, Julia Stansfield and Amy Myrbo. The *Bosmina* fossil, oospores and chironomid heads are about 0.2 mm long, the remainder are microscopic and about 50 μm in length or diameter.

Together with experience, these measurements may start to make sense in ecological terms but modern palaeolimnologists have statistical techniques that can relate collections of remains to particular conditions derived from studies of contemporary communities and the environmental conditions in which they live. The problem remains that what is preserved in sediments is selective. Soft bodies and many microorganisms leave no obvious trace and plants decompose to different extents. Increasingly, biochemical remains are used and increasingly it is possible to use fossil DNA to give a fuller picture, but for the present a degree of imagination and talent for detective work remain important.

Early in the history of freshwater science, it was recognised that water chemistry was an important determinant of what went on in ponds and lakes, both in the nature of the communities and in their productivity (see Chapter 5). The idea grew that lakes began in deep basins, with low concentrations of nutrients and ions, sparse plant populations, and clear, well oxygenated waters that supported particular genera of chironomid flies in their sediments. These conditions were called oligotrophic. As they aged, the lakes were supposed to fill rapidly with sediment and to accumulate nutrients, denser plant communities in the shallowing water and, because of their increased production, sediment that was deoxygenated at the surface in summer and supported different genera of flies. The lake thus had become eutrophic and eventually filled in to become a fen or bog and then woodland, as discussed earlier.

However, the transition from oligotrophy to eutrophy is a fiction. Lakes do not become more fertile as they age, unless the vegetation of their catchment is destroyed for agriculture or human settlement. In fact, many natural lakes probably become less productive as they age, if the climate remains steady. When they first formed they were surrounded by newly abraded rock debris from the passage of the glaciers in much of the temperate world (or perhaps by fresh volcanic ash in the tropics). Fresh rock debris has an abundant supply of minerals that are readily leached by the rain and so analyses of sediments from such lakes show great enrichment of the sediment by ions like calcium and magnesium in the first few thousands of years of their existence. This can be demonstrated also by analysing the waters in areas like Alaska where glaciers have been steadily retreating over the historic period and where newly

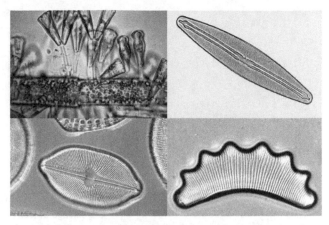

**Fig. 6.3** Diatoms are very useful in determining lake history from sediment cores because of their generally good preservation and because they come in many forms and their ecology is well understood. Living diatoms (*Gomphonema* species) that grow on plant or filamentous algal surfaces are shown top left. Bottom left is a common genus (*Navicula*) found on sediments. *Frustulia* (top right) and *Eunotia* (bottom right) generally indicate acidic, low nutrient conditions. The lengths of these cells are between 15 and 100 μm. All but the *Gomphonema* have been treated with hydrogen peroxide or nitric acid to remove organic matter and show the siliceous cell wall.

exposed lake basins of known age can be examined. The water is at first rich in major ions, but over a few hundred years becomes much less concentrated as the rock debris is leached out and ions are bound into vegetation as forests develop on the surrounding catchment. The early phase in sediment cores is associated with diatoms of eutrophic conditions, like *Epithemia*, *Gomphonema* and *Fragilaria*, not those of oligotrophic conditions, like *Frustulia* and *Eunotia* (Fig. 6.3), which develop later. The story is the reverse of what was once generally believed. But it is even stranger. Examination of the contemporary Alaskan lakes shows that the water retains about the same phosphorus concentration as the lake ages. There is a small but significant decrease but it is not great and the nitrogen content increases and then soon flattens out. This is not consistent with the transition from distinctly eutrophic diatoms to oligotrophic ones shown by the diatoms in sediment cores. Something else was happening in the post-glacial lakes that we are not seeing in the recent Alaskan ones.

That difference probably concerns the large grazing mammals that were abundant in the early millennia after the ice retreated. Such herds move around the landscape but are inevitably drawn to water for they must drink

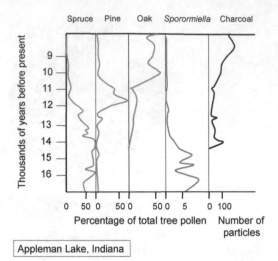

Spruce   Pine   Oak   *Sporormiella*   Charcoal

Appleman Lake, Indiana

**Fig. 6.4** From late in the Pleistocene and early in the Holocene it is common to find evidence of large herds of grazing mammals, reflected in the numbers of spores of a fungus, *Sporormiella* that colonizes their dung. At the same time there was a succession of forest types, reflected in their pollen, as climate warmed. When the mammals were hunted out, grassy vegetation built up and burned in the summer, producing ash and charcoal. Based on Gill *et al.* (2009).

copious amounts every day. And as they move, they produce dung rich in nutrients derived from the land vegetation and deposit more of it in the vicinities of ponds, lakes and streams than elsewhere. Currently, the only places where herds even approximating those of the early post-glacial period are found in southern and eastern Africa, and small lakes there show evidence of a eutrophic state related to the influence of hippopotami, elephants, buffalo, zebra and antelope. In the post-glacial period, abundance of bones in the floodplain sediments of northern Europe and in cave deposits, the evidence of cave paintings, and palaeolimnological evidence suggest that large herds were present and likely had similar influences on every continent. In Europe we had mammoths, straight-toothed elephants, aurochs, rhinoceros, wild horses and deer. The palaeolimnological evidence comes from the abundance in sediments from this period of the spores of a fungus, *Sporormiella* (Fig. 6.2), which are very distinctive. *Sporormiella* grows exclusively on wild herbivore dung. After the first few thousand years after the glaciers retreated, the spore numbers decline in the sediments and the amount of fine charcoal fragments increases (Fig. 6.4). Much evidence suggests that the herds were hunted out as people invaded the areas where they had grazed. Mammoths disappeared;

other species hung on in smaller numbers until conversion to agriculture, pasturage or forestry finally put paid to them all, except in tiny reserves. The charcoal fragments come from fires that burnt as the biomass of vegetation increased, unchecked by grazing.

The history of our lakes and ponds was therefore influenced by human activity from very early on. Were the grazer herds still with us, these water bodies would probably have begun in a eutrophic state and continued as such as the mammals delivered nutrients daily to them. Without the mammals they became less nutrient-rich with time, until the wheel turned again and crop agriculture, with use of fertilisers, and stock-keeping in the last few millennia, have reversed the trend and reinstated a new eutrophic state, albeit one that is probably far more extreme than that maintained naturally by the mammoths, aurochs and horses.

## 6.3 Back to the buckets

But all of that, no matter how interesting, is far removed from colonisation of your artificial bucket ponds in the garden. It does however, suggest an additional series of experiments where you introduce very small quantities of cow or horse dung and compare development of the systems with controls to which you add none. The quantities concerned should be very low and ideally from organic farms because dung otherwise may contain pharmaceuticals that affect colonising invertebrates. We do not know exactly what the amounts were in the post-glacial period, but about 1 gram to ten litres of water might be about right. You could, of course experiment with several different loadings, but then you may be running out of lawn space.

In time, a surprising variety of invertebrates and algae will find their way into your ponds. Amphibians may also invade and pose the problem that they may enter some of your replicates but not others. Amphibians are voracious feeders and your ponds will be too small for even one frog, toad or newt to be able to persist without depleting the pond of most of its animals. You have to make the decision whether to remove the amphibians to a convenient larger pond or accept their presence and the variability it will bring. That sort of decision has frequently to be made by working scientists. Your ponds will undergo a succession, though not the hydrosere in which they become encroached upon by plants and succeed towards a fen, bog or wet woodland, of course, but you can study those aspects in a less controlled

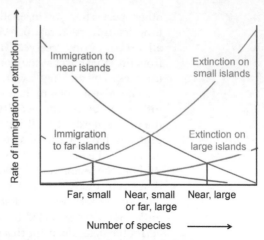

**Fig. 6.5** The essence of island biogeography theory. Immigration rate to an island is greater the fewer species that are already there, so that more habitat space is available but is lower the more distant the island from the source of species on the mainland, because of the greater difficulties of reaching it. Once on the island, extinction rate increases the more species are packed in, and is greater on small islands where competition is greatest. The 'equilibrium' number of species is given where the immigration and extinction curves cross. It is lowest for distant small islands and greatest for large islands close to the mainland.

way by finding an old pond and looking at its sediments, or a sequence of ponds of different ages. The best sources of these at present are abandoned gravel pits, which are often now turned over to conservation organisations. Often they are then managed to favour particular states or stocked with fish and the natural successional sequence is interfered with, but there will still be an interesting history to be uncovered.

## 6.4 Island biogeography

The second aspect of theoretical ecology that can be investigated using ponds, either in buckets or in the landscape, is that of biodiversity. In a famous book published in 1967, Edward Wilson and Robert Macarthur developed a theory to explain the richness of organisms on islands in the ocean (Fig. 6.5). Ponds can be treated as islands also, but islands of water in a sea of land. The comparison is not exact because ponds are connected through the network of ground and surface waters, whilst oceanic islands truly are isolated from other land. New oceanic islands (created by volcanic activity for example) must be colonised from the continental land masses and Wilson and Macarthur suggested that there was a balance between the rate of arrival of new species and the rate at which they became extinct on the islands. This was because they could not survive for a variety of reasons, including unsuitable habitat and competition with already established species. With time the number of species arriving would cumulatively increase. The rate of extinction would be low at first but would increase as more and more species established. The point at which the two curves intersected (Fig. 6.5) would represent an equilibrium that would describe the number of species on the island, but of itself would not explain it. The explanation would lie in two independent factors: the size of the island and the distance from the mainland and the source of new species. The greater the distance, the lower would be the colonisation rate, whilst most risks of extinction would be just as high as on an island near the coast. The equilibrium point would thus be at lower numbers of species with distance from the mainland. But the bigger the island, the more varied it would likely be, and therefore the greater the variety of organisms it could support, whilst the possibilities of refuge against predation, or of finding slightly different niches in which to avoid competition, would be greater. Extinction rates would be lower on large islands. The point of equilibrium would thus be skewed towards a greater number of species with increasing size of island.

Wilson and Macarthur were able to show that these principles were valid for individual groups of species, like reptiles and amphibians on the archipelago of the Caribbean Islands, and for birds in Malaysia. Later the distance effect was shown on small islands of mangrove that had been cleared of their existing insect fauna by fumigation. The size effect is more difficult to test because it requires a very large range of sizes, over several orders of magnitude. In the Caribbean, some islands, like Redonda were only 1 km² in area, whereas at the upper end, Cuba has an area of 110,000 km². There is also a corollary concerning the turnover of species. The theory predicts that there should be a continual replacement of species with time as new ones arrive and existing ones disappear. The difficulty of testing this is that the life spans of investigating humans may not be long enough for a sustained study. Trees, in particular, may live much longer.

Theories are intended to simplify complex situations so as to improve understanding, but reality adds an interesting dimension. The theory of island biogeography works well with truly isolated islands, but it has been applied to less isolated situations, such as mountain tops (where it works well for mammals but less well for birds) and patches of remaining forest in seas of developed land. In these there is a process first of loss of species because the former diversity included species that require large territories, and species that are highly specific for niches that might no longer be left in fragments of the original. The settled state that is eventually studied will be affected by this history. The predictions of the theory might be met if the patches are truly isolated and not connected by corridors of hedgerows, or stepping stones of habitat that can at least temporarily accommodate species on the move. It is a theory that has been applied to optimise the design and number of nature reserves needed to conserve the maximum biodiversity.

The interesting question is whether the theory holds for ponds. If it does, then there would be implications for pond restoration and conservation. For example bigger ponds should be better than smaller ones, and ponds grouped together near existing ponds and connected by wetland easily traversable by colonising species should give maximal biodiversity. Your experimental bucket ponds can be used to determine colonisation and extinction rates. In a sea of lawn they approximate to islands quite well. You can also investigate turnover because the algae and invertebrates have short life histories. Does the community remain stable once established or do the species change frequently from year to year? This has implications for conservation, which in the UK is much devoted to maintaining particular species.

You can't do much about testing the area effect with the buckets, but you can certainly do it using existing ponds and lakes of as great a size range as can be mustered. It is probably best to consider only some groups of organisms. Plants and birds lend themselves well and also particular groups of invertebrates such as dragonflies and damselflies, and snails, where the taxonomy is relatively easy and there is a reasonable number of species. Remember, though, that other factors, like the water chemistry, can have strong influences and so select your sites from as uniform an area as possible. There is now a whole branch of statistics called multivariate analysis that can separate out (to some extent) the influences of many different variables acting at the same time.

Island biogeography theory gives some insights into what determines biodiversity, but answers the questions only at a very general level. Can more detailed questions be asked? There are some general ideas that can be tested. First, exactly what is meant by biodiversity and how can you measure it? Secondly, why do some habitats have a naturally low biodiversity? Are they inferior to ones where biodiversity is high, the so-called hotspots? Thirdly, do habitats have species that are redundant; is the biodiversity greater than is needed for the functioning of the ecosystem? The concepts of biodiversity are bound up with those of ecological niche and there is much that is unknown about the limits of niches for pond organisms.

## 6.5 Biodiversity measurement

To take these issues in order, what is biodiversity? Essentially it is variety, but of what? At its simplest, it is the number of species, and there is little problem if the count is confined to some distinctive group, like birds or dragonflies. Such a count is strictly of species richness, however, not diversity. Diversity has to take into account relative numbers, so that if a pond receives 1,000 visits among 10 species of birds in a year, it has higher diversity if each of the species visits 100 times than if one species visits on 910 occasions and the other nine on 10 occasions each. There are mathematical indices that take account of both species number and relative abundance, for example the Shannon-Weaver Index where:

$$\text{Diversity} = -\sum p_i \log p_i$$

$p_i$ is the proportion of each species in the total count of all species and the logarithm should be to base 2, although a set of estimates will bear the same ranking no matter what base is used; 10 or $e$ (= 2.718), the base of natural logarithms, are frequently used. But base 2 is fundamental because it builds in the concept of two states, presence or absence, whereas other bases do not incorporate this meaning. Using base 2, highly diverse communities might have values of 3 or 4, poorly diverse ones <1 and to convert values in $\log_{10}$ or natural logarithms to those in $\log_2$ the equations are:

$$\log_2 x = \log_{10} x \div \log_{10} 2$$

$$\log_2 x = \log_e x \div \log_e 2$$

Diversity in base 2
    = diversity in base 10 ÷ 0.301, where $\log_{10} 2 = 0.301$

Diversity in base 2
    = diversity in base $e$ ÷ 0.693, where $\log_e 2 = 0.693$

In our case of 1,000 bird visits, the two respective diversities in base 2 would be 3.32 for equal numbers of visits by the 10 species and 0.70 where one bird dominates. The values in base 10 would be 0.1 and 0.021 and in natural logarithms 2.3 and 0.49. The relative magnitudes stay the same irrespective of base used but the absolute numbers change and it is important in making comparisons across different investigations to understand this.

It is easy to calculate simple numbers, but more difficult to know what they mean ecologically. Obviously a lot depends on what is counted. Species seems an obvious unit, but for many groups, not least the microorganisms, it is difficult to delineate a species because genes are continuously exchanged in successive generations by a simple injection process. In algae and protozoans there is great variation within what are thought to be species, but in which the necessary demonstration, for the definition of a species, of ability to produce fertile offspring in a breeding experiment is often not available. Growing algae in different conditions of water chemistry or in the presence of chemicals from grazers can give very different forms that might be considered different species were they mammals or birds.

The same sort of variation occurs in invertebrates and even fish differ a great deal among individuals. Fish indeed have personalities; some individuals are timid, others bold and sometimes they have their own individual feeding preferences. Cryptic diversity, where populations that look much the same in different places have very different genomes (sets of genes), appears to be common. If biodiversity is to be an important measure in ecosystems, it should be total biodiversity that is measured, and that probably means diversity at the level of the genes, but that is presently impossible as a practical consideration. We know the base sequences of DNA from an increasing number of organisms, but it is not yet often possible to define what is a gene among these sequences.

We have a general understanding that the degree of variety must mean something, but usually all we can do is to count species and individuals for well-known groups and assume that the group is representative for the whole system. You can test this by taking a series of ponds and measuring the species richness and some measure of relative abundance of, say, vascular plants, birds, snails, dragonflies, and mayflies, all of which are easily identified and where you stand a good chance over a year or so of drawing up a comprehensive list. You can then rank the ponds by number of species and by Shannon-Weaver diversity index for each group. Do you get the same rankings of diversity in the ponds for each group? It is an important question because conservation assessments, and decisions to buy this patch of land or that, are often based on only one or two groups and then usually not the microorganisms, algae or invertebrates. At the same time you can measure aspects of the ponds that might determine diversity, which brings us on to the second question posed above. What does determine biodiversity?

## 6.6 What determines biodiversity

Think of a temporary pond, essentially a puddle. So long as it lasts a few weeks it will not be devoid of life. If you can find some dried-up ponds, take some of the dry soil and place it in a jam jar with rainwater that you have boiled and cooled (to remove any organisms that would have been carried in the rain). Almost certainly some organisms will hatch out within a few days. They might be algae, protozoans, crustaceans, tardigrades or fly larvae, and sometimes they will be very abundant (and feeding on organic detritus or algae that grow on the mud), but there will be only a few species, almost certainly fewer than ten animals. In contrast, a permanent pond in the same general area will have up to a thousand, perhaps twice as many. In one study of Priest Pot, a pond at the head of Esthwaite in the Lake District an incomplete list (for certain groups were not examined thoroughly or at all), revealed about 870 species.

The difference between the temporary and permanent pond is in the predictability of the habitat. For similar reasons a lake will have fewer species than a comparable area of ocean, because in geological terms a lake is an ephemeral, unpredictable habitat compared with the ocean. It might dry up, or be washed out by floods, or freeze severely from time to time over a few thousand years. The ocean, because of its size, will not. Likewise, a rocky mountaintop will be less diverse than a tundra on its exposed lower slopes and that less diverse than a forest in a more sheltered valley at comparable altitude. Predictability of habitat means that the organisms that live there will experience steady conditions and therefore will be able to evolve to be more specialised than in an unpredictable one, where physical conditions and availability of food might change enormously from month to month or year to year. Such denizens will need to be very flexible generalists to survive. Our general understanding is

**denizen**
an organism that lives in or often is found in a particular habitat

that in an unstable habitat (the early phase of a succession is one such) the niches that organisms can occupy are fewer than in a stable one (like the mature phase). Likewise the bare open plastic bowl with nothing but water and sand is very different from one that has accumulated some detritus and some structure from colonising algae or even plants. I once set some plastic bowls, filled with lake sediment from a lake, back in the lake bottom, surrounded by exactly the same sediment. They rapidly acquired a more diverse fauna than the surrounding sediment away from the bowls. The plastic rims had provided a new structure.

Structure gives all sorts of possibilities for increasing the number of species. Try putting some extra structure in a pond that does not have much. You can use bamboo canes pushed into the bottom in a 5-cm spaced grid, or scrunched up fruit cage netting, or the sorts of bricks that have holes in them, and compare what colonises these structures with what is present in the surrounding sediment. You can also use artificial plants. Strands of polypropylene rope sown into a piece of sacking or tied onto a plastic grid will do quite well. You can buy plastic plants that are made for aquaria and resemble the different structures of real plants. Fix them into mats of area about 30 cm x 30 cm or more, weight them, and lay them onto a pond bottom. Remember to incorporate some means by which you can find them again. Recover them a few weeks later by inserting a pond net underneath them. Is there a relationship between complexity of structure and the eventual invertebrate diversity? As always, randomise their position and replicate the treatments and controls and you can use simple statistics to find out what your results mean.

**control**
a standard (usually untreated) for comparison with the treatments in an experiment

Niche is not just a physical concept however. An organism's niche also includes aspects of its feeding, behaviour, seasonality and life history and there are two ways of looking at it: the fundamental niche and the realised niche (Fig. 6.6). The fundamental niche describes all the possible conditions under which an organism can survive and reproduce; in other words, persist as a species. There will be a huge number of combinations of potential conditions and indeed no one has yet accurately described such a niche, for we do not know the complete list of conditions for any species. It is a theoretical concept and we talk of the set of conditions as the niche hyperspace, but it is a space that has many dimensions (as opposed to the three dimensions of conventional spaces, or four if time is included) but it helps to understand that a species can

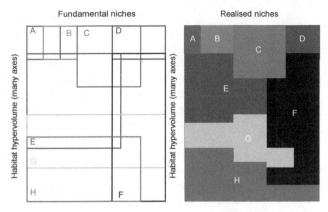

**Fig. 6.6** There are many variables that determine the possibilities for living in a particular habitat. They may be physical (weather, type of sediment, tree debris, rocks), chemical (thousands of different substances dissolved in water or forming sediment, or biological (competition, parasitism, predation, symbiosis). There are thus many dimensions to a habitat and it is impossible to express it in the three possible on a printed page. We thus imagine the vast number of dimensions flattened onto a page and call it a hypervolume. Individual species have tolerances to all of these variables to a greater or lesser, complete or zero, extent and this envelope is called the fundamental niche. The left-hand diagram shows the fundamental niches of eight species, A-H. The tolerances overlap and this means that there will be competition for the hyperspace that is common to the two. The outcome is that only one species can occupy any given part of the space, so that the realised niches (right-hand side) do not overlap. There is, however, potential continual competition where two realised niches abut if the boundary lies within the fundamental niches of the two. Natural selection tends to minimise this competition through greater specialisation and contraction of the realised niche away from the boundaries. One effect of this is that a new species, moving in from elsewhere, will attempt to occupy the space opened up.

persist only if its fundamental niche space overlaps with the set of conditions, the habitat hyperspace, pertaining in a particular place (or places, for some species move or migrate seasonally).

The fundamental niche spaces of similar species will often overlap. Experiments using simple systems with *Paramecium* species, which are ciliate protozoans, and with flour beetles, have shown that two different species cannot indefinitely occupy exactly the same conditions. One of the two species always proves slightly fitter to occupy the conditions and the second species reproduces less well and eventually disappears in a process called competitive exclusion. If they are to coexist they must each retreat to a part of their fundamental niche that does not overlap, or they must evolve so that their niche hyperspace changes so that parts do not overlap with other species. We talk of their realised niche as the set of conditions to which this process forces them to retreat. The habitat hyperspace is then filled with the realised niches of a set of species whose niche boundaries do not overlap. Initially this may be accomplished by competition and change in behaviour. Eventually the continuous operation of natural selection will result in each species contracting away from the edges of their realised niches so as to minimise competition with other species (for competition is wasteful and reduces the ability to survive). Sometimes this process will result in a new species being able to invade the marginal space and sometimes in differentiation of two new species occupying a former realised niche that only the ancestral one had occupied before. In doing so they will be occupying two new and smaller realised niches.

Unpredictable habitats have species that must have large realised niches to allow for the vicissitudes of change. Such species will be very flexible, have broad diets, invest much energy in reproduction (for they may be easily eliminated by extreme conditions) rather than defence against predators, and have short, simple life histories that do not depend on other species for their completion. They will be the sort of species that are pioneer colonists or those that can survive in impermanent habitats like temporary ponds, and streams that dry up frequently. In a sense the entire freshwater biota has some of these characteristics, for all freshwaters are ephemeral on even the shorter geological time scales. Predictable habitats will have species with the converse characteristics: the realised niches will be small, with extreme specialisation in conditions and diet, long life histories with an emphasis on long-term survival through evolution of defences against predators, and frequent collaborative symbioses with other species. In freshwaters these characteristics will be likely to be found in well-structured habitats such as ponds with well-established vegetation, reliable water supply and surroundings that have natural vegetation rather than the fluctuating land use where there is much human activity.

## 6.7 Niches in ponds
Much work has been carried out in listing the species found in ponds, much less in exploring relationships

that throw light on the separation of their niches among species. Experienced naturalists will often gain a feel for where particular species will be found without being able to define precisely what the conditions are. Refining these feelings offers many opportunities and lends itself particularly to plants and invertebrates. To build up a picture of niche separation, the key is to choose appropriate species carefully and to find a limited number that are occurring together in a pond, or better, in several ponds. Snails, the larger crustaceans like *Asellus* and *Gammarus*, mayflies, dragonflies and damselflies and flatworms are ideal among animals, emergent plants, pondweeds (*Potamogeton* species) and duckweeds among plants. It is very important that you are sure that you can accurately identify the particular species you choose, and are aware of local variation (which is common). The first step is to record exactly where you find them in the ponds (you will need to make a careful map), with respect to water depth, adjacent plants and associated other animals, and bottom substratum, and when you find them. You need to sample throughout at least a year. For insects, remember that important parts of the life history are during emergence, mating flights and egg laying.

You can then test ideas about how related species separate their niches by experimental work indoors. One of the great advantages of working with freshwater animals is that they often take readily to life in an aquarium (especially compared with marine species which are much more fussy). Keep your animals in an aquarium (any glass bowl will do: large Pyrex casserole dishes or cooking trays from the kitchen, for example) filled with water from the pond and with sections of the bottom filled with different substrata (leaves, mud, sand, waterlogged twigs) and watch patiently. If the area of water surface is large relative to the volume, and the temperature is not high, you will not need artificial aeration, but use an aquarium bubbler if you have one. You may find that the animals move to particular substrata or that they are active during particular hours of the day (or night). Make observations of your chosen species, singly at first and then in pairs to see whether their behaviour is changed in the presence of each other. Try adding predators like dragonfly larvae, leeches or sticklebacks. Gradually, from your field observations and your kitchen laboratory you will build up a picture that will give you insights into how niches are separated, though it will be a very much simpler picture than the reality.

## 6.8 Redundant diversity?

A big question about diversity concerns its very existence. For most of Earth's history, when there was life at all, from about 3.8 billion years ago until just over half a billion years ago, there were no multicellular plants and animals. The biosphere was maintained and became steadily more oxygenated by microorganisms. Only in the last 540 million years or so have any of the animals and plants, to which we now devote a great deal of effort to conserving, been present. We have little clue as to whether the overall biodiversity under the microbially dominated state was greater or lesser than under recent conditions but there does appear to be a redundancy in our present biodiversity. For example a pond might have twenty species of plants, up to a thousand algae and protozoans, more than a hundred macroinvertebrates and the visitations of twenty or thirty birds and mammals. It has many fewer basic functions to maintain: decomposition of organic matter coming in from the catchment; photosynthesis; conversion of the produced matter through consumption and decomposition back into materials that can be recycled. Ultimately the operation of the biosphere amounts to the efficient recycling of carbon in a watery medium. So why are there so many species, both collectively and locally? A pond might function perfectly well with one emergent plant species and one submerged, a handful of decomposer and nitrogen fixing bacteria, a few species of algae broadly adapted to cope with the changing seasons, a shredder or two and a couple of deposit feeders and predators. There is no obvious need for birds or mammals. It is indeed possible to maintain in the laboratory self-regulating systems with very few species. Neglected freshwater aquaria are good examples.

On the other hand, increasing numbers of studies are showing that slight differences in niche among the apparently redundant participants lead to greater efficiency in the carrying out of each process. Five different shredders process leaf material more rapidly than an equivalent biomass of only one. Fifty algae of different pigment compositions make fuller use of available light in water that warms from 3°C in winter to 23°C in summer. Again it is possible to investigate this using simple equipment. Discs of fallen leaves, 2.5 cm across, cut with scissors or a small pastry cutter, which have been left in the pond for a couple of weeks to acquire their hyphomycete fungi can be placed in jars with equal biomasses or numbers of one, two or three shredder species. The progress of the shredding can be assessed from the area or weight of leaf left. The process of natural selection in tending to avoid competition between species by contracting their niches away from the boundaries will inevitably tend to increase the number of species and therefore biodiversity. It may also, inadvertently, improve the chances of survival of the system should conditions change rapidly and markedly. 'Redundant' biodiversity may be an insurance policy against change. There will be replacement species that can take over in the new conditions if the former dominants do not cope. The more species available, the greater is the spread of the risk. Sometimes this is known as the rivet hypothesis: the more rivets that are used to join together the steel plates of a ship's hull or an aeroplane, the less likely it is to fall apart in a storm. Try simulating this in kitchen aquaria, using a set of aquaria that you have established and maintained as closely similar as possible over a few weeks by mixing water and sediment among them so that they have become uniform. Increase the temperature of some of the replicates, by say 5°C using an aquarium heating system as used for maintaining tropical aquaria. How does the system change? Are its functions intact? Or does everything die?

A great deal of effort is spent by ecologists trying to convince politicians that our current huge loss of biodiversity is important. Many species are comparatively scarce, or live in very limited niches and loss of one or two would undoubtedly make very little difference to maintenance of the biosphere. But loss of more than a few means the tearing up of the insurance policy. The great American ecologist, Aldo Leopold, wisely observed:

*The last word in ignorance is the man who says of an animal or plant, "What good is it?" If the land mechanism as a whole is good, then every part is good, whether we understand it or not. If the biota, in the course of aeons, has built something we like but do not understand, then who but a fool would discard seemingly useless parts? To keep every cog and wheel is the first precaution of intelligent tinkering.*

# 7 Food webs and structures in ponds

Fish are the animals most people first think of, when freshwaters are mentioned. Trying to catch tiddlers was a favourite occupation, now sadly discouraged by a generation of parents made fearful of almost everything, of my childhood. Angling, however is not to be suppressed and is widely practised. In warmer countries and eastern Europe, freshwater fish are major sources of protein and fishponds are serious business. In garden ponds, goldfish and their expensive relatives, koi carp are much admired. The more and the bigger the fish, it is widely believed, the better. But that is not necessarily so, especially for pond conservation, and large numbers of big fish are usually only a dream for fish farmers and competitive anglers. Fish are very important in freshwater systems but in much subtler ways than usually understood. It is time to look at food webs in ponds.

The idea of food chains is quite old and the melding of a set of chains into a web dates back well into the twentieth century. The web starts with a source of energy, always (except in deep ocean vents) the sun, and the raw materials of carbon dioxide, water and the necessary mineral nutrients. Plants and algae are the primary producers, whether in the water or as material washed in from the land and are eaten by grazers (herbivores) or detritus feeders, and those in turn by primary then secondary consumers, and possibly further levels where production is very high. It is a very familiar concept summed up biblically as *all flesh is grass*. But matters are not quite so simple. Different plants and algae produce material of variable edibility. Cyanobacteria, for example, have rather tough walls and may produce toxins, which make them less attractive. Plants are bulky, rich in cellulose but not protein and full of air spaces and may be shunned in favour of the periphyton on their surfaces. Grazers may be adept at finding refuges against being consumed and consumers against other predators. Material that is eaten may not be entirely digestible and is defaecated, perhaps to be colonised by bacteria that are then eaten by protozoans and small animals. Simple chains with single species at each level do not exist. Almost every animal in freshwaters takes a variety of food and is essentially an omnivore but with its own preferences. There may be cannibalism, especially among fish, of adults on younger stages. The freshwater food webs, perhaps because of the unpredictability of freshwater habitats, generally lack the highly specific feeding relationships that are found in forests, particularly among insects.

Food webs used to be constructed from detailed observations on what was observed to eat what, and from dissections of the guts of the animals to discover what they had eaten. The problem with that was that much of what was being eaten might be eaten at night and therefore unobserved, and that the softer prey was being digested and therefore made unidentifiable. Nonetheless it was possible to discover a great deal from the less digestible parts, such as wing covers, head capsules, legs and carapaces. Matters are now more convenient because measurements of stable isotopes of carbon and nitrogen can be used to determine the origins of food that has been incorporated into the body and the position of each organism along the food chain.

The methods depend on taking single animals, or pieces of plant, or small samples of detritus or sediment, and burning them to vapourise their carbon and nitrogen. The carbon and nitrogen oxides are then passed through a mass spectrometer, coupled with a gas chromatograph, which can measure the amounts of different isotopes of carbon and nitrogen. Carbon isotopes have been discussed earlier and can tell us whether the material originated through plants photosynthesising using atmospheric carbon dioxide (land plants and emergent reeds) or underwater plants using carbon originating from bicarbonate and the rocks of the catchment. Plants using carbon dioxide from the air tend

$^{12}C$ and $^{13}C$
isotopes (forms) of
carbon with respectively
12 and 13 protons and
neutrons in each atom.
pronounced carbon-12
and carbon-13

to be low in $^{13}C$ because it is heavier than $^{12}C$ and diffuses into the plants less rapidly. On the other hand plants taking up bicarbonate, or carbon dioxide generated under water from it, tend to have a little more $^{13}C$, because rocks from which the bicarbonate is derived are richer in $^{13}C$. The $^{13}C$ is most conveniently measured relative to a standard and the results expressed as a deviation ($\partial^{13}C$) from the standard (a rock fossil known as Pee Dee Belemnite) first used for the technique. Thus material with a very negative $\partial^{13}C$, perhaps −30 parts per thousand (ppt) or so, tends to have come from plants using air as a source of carbon dioxide, whilst those using bicarbonate have a deviation, perhaps −20 ppt, that is smaller. The carbon isotope can thus give us an idea of the source of organic matter on which an animal has fed, and in freshwaters can tell us whether it has been derived from material produced in the water, or derived from washed-in land vegetation or the emergent reed vegetation. As yet we

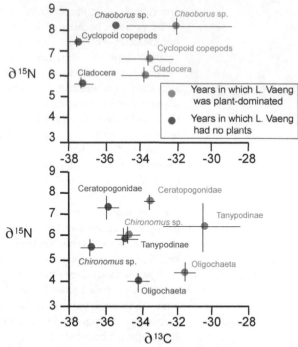

**Fig. 7.1** Analysis of the stable isotopes $^{15}$N and $^{13}$C in samples from a pond can give information on the food sources and food webs. The greater the $^{15}$N contents, the higher in the food web the organism is feeding. The more negative the $^{13}$C content, the more likely it is that the organism is dependent on organic material coming in from the land. In Lake Vaeng, Denmark, there are years when the lake has an extensive coverage of submerged plants (green dots and labels) and others in which it completely lacks plants (brown dots and labels). In the latter years the animals of both the zooplankton (upper) and sediment (lower) depend on incoming organic matter. Based on Boll *et al.* (2012).

cannot distinguish between reedswamp and truly land vegetation with this technique.

The nitrogen isotope gives us different information. $^{15}$N, a scarce isotope compared with the more usual $^{14}$N, becomes enriched in a relatively steady way by 3–4 times at each step in the food chain. The change in $^{15}$N is also expressed as a change, $\partial^{15}$N parts per thousand, relative to the $^{15}$N content of the atmosphere. By graphing the $\partial^{13}$C, as the horizontal (*x*) axis, and $\partial^{15}$N, as the vertical (*y*) axis, for each sample, a useful picture is obtained (Fig. 7.1). Carnivores will have high enrichments of $^{15}$N, and lie high on the *y* axis, detritus feeders and plant grazers much lower ones, towards the bottom of the axis, and plants and algae lowest of all. Water plants will lie to the right of the *x* axis, land

**Fig. 7.2** The larva of the caddisfly *Tinodes waeneri* builds cases, which catch small invertebrates, on stones and cultivates algae on the case material, which it then eats. Eventually it hatches to an adult (inset). Photographs of adult by Graham Callow and larval cases by Jacob Forster.

plants and detritus derived from them towards the left. Animals feeding on water plants will be to the right, those feeding on imported detritus to the left. Many animals will have mixed diets and lie intermediately, but mathematical techniques can be used to calculate the relative amounts of each component that they take, given the isotope contents of the original foods.

The picture obtained is that most animals in waters and mires are omnivores with some degree of preference for one food source or another. Some caddisfly larvae, for example, living on underwater rocks, spin bags of silk, in which they live but also use to catch smaller prey animals that wander or drift in. But not only do they create a home and larder in their bags, they incidentally cultivate algae that grow on the bag, fertilised by their oozing excretions. They eat the algal-infested bag material at one end and make new bag material at the other: a sort of mobile home and garden (Fig. 7.2).

## 7.1 Population changes

Food webs are central to understanding how ecosystems function. Through the relationships of what eats what, come ideas about population numbers of organisms and how these numbers are controlled by the amounts of food available and by predation. Of course these are not the only influences on numbers; environmental changes such as floods and droughts, inclement weather, deoxygenation and nutrient shortage all contribute. We can distinguish density-dependent control in the case of predators and their prey and density independent control in the case of environmental changes. One basis of density dependent control is that as numbers of prey rise, there will be a corresponding increase in the number and activity of predators, which will then reduce the number of prey until such a point that the predator runs out of food and declines, allowing the prey to increase again, and so forth. Theoretically this leads to cycles of numbers of predator and prey, the two graphs being slightly offset from one another. Similarly growth of algae can be determined by uptake of nutrients to a point of depletion, when the population falls, followed by recovery when nutrients are replenished. Another basis, especially in birds and mammals is through the establishment of breeding territories and social behaviour. A territory is an area that can provide cover and food for a pair and their offspring. The land can provide a finite number of such territories, so individuals not strong enough to hold a territory are excluded from favourable habitat and become more vulnerable to predators. The population size is thus regulated by the availability of territories, which determines the density of animals.

Density independent effects tend to be much more random (Fig.7.3) and lead to graphs of numbers that are

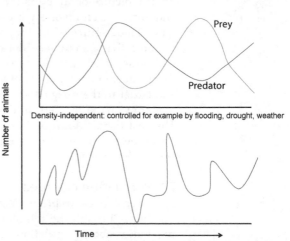

Density–dependent: controlled for example by predators or availability of suitable habitat

Prey

Predator

Number of animals

Density-independent: controlled for example by flooding, drought, weather

Time

**Fig. 7.3** Idealised curves for populations controlled by their density or by external factors. Often there may be phases of density-dependent control interrupted by phases of density independence, following, for example a major flood.

spiky and change greatly from year to year. And super-imposed on density-dependent and density-independent changes are life history changes. A given animal may start with low numbers in late winter, then build up its numbers until a point is reached where predators are attracted or it reproduces and the adults die to be replaced by eggs that overwinter awaiting hatch in the spring. If a year is relatively cool there may be an influence of environment that delays reproduction and perhaps extends the life history through to a second year. For many insects there may be one or two years of development as larvae, with there being fewer but larger larvae at every instar until the insect emerges as an adult causing the aquatic population to decline to zero and there being a replacement with eggs.

Following and explaining the changes in numbers of an animal in a pond is thus a challenging, but interesting business. It is worth tackling though, for the insights it can give into the ecology of particular organisms. Choose a common animal: snails and pond shrimps are ideal rather than unusual or charismatic ones. Rarities tell us much less than those animals that are the bread and butter of ponds. Find out from reference books as much as you can about your chosen species before you start. You then need to devise a sampling procedure that is as unobtrusive as possible and disturbs the habitat very little. It need not give absolute numbers (per unit area for example) but it does need to be applied consistently and relatively frequently. Sweeping through with a pond net has its problems (though it is often used) but placing of artificial substrata that can be replicated and replaced, is much more sensible. Clumps of polypropylene rope, scrunches of nylon netting, bristle brushes, even lavatory brushes might be used. You need to leave them undisturbed for at least three times the period between samplings for recolonisation so if you take five samples on each occasion (replication in everything is the rule), then replace the washed substrata, you will need twenty substrata (fifteen for sampling, so that each individual substratum is only sampled on every fourth sampling occasion, five for contingencies like disturbance or loss) to be placed in the pond. Time between samplings depends on the animal, its abundance and ability to recolonise the substrata, but three weeks or a month will often be about right.

You need to be able to relocate them and take them out with minimal disturbance. After sampling they need to be thoroughly washed out in a bucket to remove all

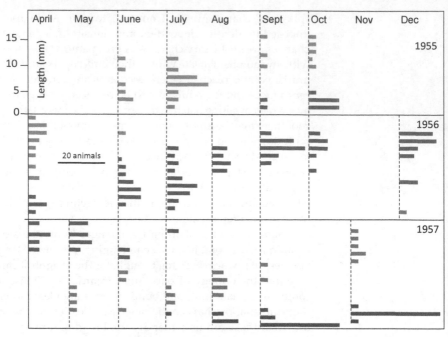

**Fig. 7.4** Changes in the size and numbers (per standard sweep net sampling through the vegetation) of the nymphs of the mayfly *Pyrrhosoma nymphula* in Hodson's Tarn, English Lake District. Different colours show different generations (cohorts). The red cohort suggests that a typical pattern is for egg laying in late summer then a year's development followed by emergence and egg-laying in the next year, with an overlap between generations. Sampling here was not regular and that leaves questions about when egg laying occurred. It may have been over an extended period as nymphs reached an appropriate size or it could have been all at one time: in August or September in 1955, probably October in 1956 and as early as July in 1957. It is difficult to catch the smaller nymphs and they may have been present earlier than detected. Based on data from Macan (1973).

the animals and your chosen species picked out, before replacement for recolonisation and sampling in the distant future. It is useful to measure the sizes of the animals by some standard method and of course to count the numbers so that you obtain a profile of numbers for different sizes (Fig. 7.4). Note also any evidence of mating (pairs of animals in precopular stages) and egg pouches and number of eggs. You may need to dissect specimens to discover this.

It is useful also to keep records of environmental factors, particularly temperature as this can greatly influence life histories. You may also wish to carry out experiments indoors with potential predators. The reference books will give you clues to likely species, and to monitor numbers of these also. In aquaria you can arrange chambers on the bottom with different potential prey and discover if particular predators (dragonfly and damselfly larvae, flatworms, beetles, bugs, sticklebacks) have preferences and you can determine the numbers eaten per unit time. It's important to provide a natural setting in the aquarium to determine this. On bare glass floors predators will find and eat anything (and everything) very quickly. Over a year, or better a longer period, you should have enough data to see whether there is a steady population or a regular cycle (determined by life history or by predation), or whether numbers are irregular and probably determined by environmental changes. Your experiments will allow you to formulate ideas about the causes of regular cycles. You will not come to final answers, but you will learn a lot about your organisms and the processes involved.

## 7.2 Stability in food webs

Theoretical ecologists have developed ideas about stability in food webs. If one item is removed from the web, how does it affect others and does it make the web less stable through problems for other organisms? Are simple webs more liable to disturbance than complex ones? What proportion of all possible linkages through a web actually occur and is there meaning in this? Despite a lot of effort and some fairly arcane terminology, especially now that these things can be simulated in computers, little that might be graced as a general theory has emerged. We are still casting around.

A few years ago, I was visiting the Museum of the Gulbenkian Foundation in Lisbon. It has a sculpture garden and, like all good gardens, it had ponds, quite a few of

**Fig. 7.5** A clear pond, lacking fish, and a turbid pond with goldfish in the Gulbenkian Museum Sculpture Gardens, Lisbon

them (Fig. 7.5). They were connected, as far as I could see, into the same water supply, though my investigations of the plumbing were ended by a security guard who did not seem to share my spirit of unfettered research. There were two sorts of ponds, one of which had goldfish, and water that was bright green from suspended algae, but no water plants, and the other which had clear water and, albeit somewhat scrappy, submerged water plants. It was, in fact a neat demonstration of a phenomenon that still has high profile in ecological research: that of alternative stable states. The concept is that under broadly similar physical and chemical conditions, there can be several markedly different biological communities. Markedly different does not mean that they differ by a few species; all communities are unique in composition when the details are explored. It means that there is a major structural difference: in this case plant dominance, with all the physical structure that plants can provide, versus a rather amorphous algal-dominated community. Other examples of alternative states include bare rocky seashores dominated by sea urchins, versus kelp beds with a rich community of invertebrates, fish and birds on Pacific shorelines of North America, and the alternatives of grassland versus forest over much of Europe. In all cases the communities can persist for a long time once established and in all cases there are biological mechanisms that maintain the differences. In ponds and shallow lakes they often depend on the fish community, on the Aleutians, the presence or absence of sea otters, which feed on sea urchins, and in Europe the maintenance of ploughing or grazing by cattle and sheep. In garden ponds and domestic aquaria a surfeit of goldfish often leads to loss of plants and dense green algal populations.

Fish thus have a key role and people often alter fish communities. Some species are removed, others, often alien species like goldfish and common carp, are stocked. That opens up in your garden pond, or experimental bucket ponds at least the possibility of some experimentation. But first, how do these alternative systems operate? There are two important ideas, those of buffer mechanisms (Fig. 7.6) and of switch mechanisms (Fig. 7.7).

## 7.3 Buffers and switches in ponds

Plant-dominated ponds remain as such, first of all, through maintaining calmer waters in the face of wind. Planktonic algae depend on wind-induced eddies to keep them in suspension. They sink to the bottom without these. Plants

**allelopathy**

the release into the environment by an organism of a chemical substance that acts as a germination or growth inhibitor to another organism

Fig. 7.6 Plant-dominated and algal-dominated waters each have buffer mechanisms that maintain the status quo.

Fig. 7.7 Some of the mechanisms that can destroy plant communities and allow a switch to algal dominance

still the water somewhat and disfavour competing phytoplankton. Then some plants produce organic chemicals that suppress algal growth. These are called allelopathic substances and we know little about their exact composition, but the garlicky smell that emanates from stoneworts may be linked to their abilities to suppress the growth of microalgae. Uptake of nutrients from the water by a dense plant growth may also help to limit algal growth, though small algae compete more effectively than large plants in nutrient uptake. More important may be that in the organic-rich, deoxygenated conditions at the bottom of the plant beds, conditions for denitrification are ideal and the algae are denied a nitrogen supply. The plants also have access to phosphorus supplies in the sediment that the algae do not.

The most subtle buffer mechanisms, however, are that plants provide refuges against fish predation for zooplankters, which move out at night to graze algae in the water, and for larger invertebrates like snails and mayfly nymphs that graze on the periphyton that covers the plant surfaces and can cut off light to the tissues below. Most fish need to see their prey and then assess whether to make an attack to catch it. In a dense plant bed, light is restricted and the clutter of leaves and stems means that the fish can see less far and are deterred by the risk of damage to their heads during an attack on a zooplankter. For the periphyton grazers, the same strictures apply, with the addition of even closer concealment in the nooks and crannies, corridors and caverns within a complex plant bed. The edges of the plant beds also provide lurking habitat for piscivorous (fish-eating) fish like pike (*Esox lucius*). Such predators reduce the numbers of small zooplanktivorous fish and in turn a large zooplankton population of efficient grazers, like the cladocerans (Chapter 4) can persist.

A plankton community, once established, has advantages over the large plants, reflected in a different set of buffer mechanisms. First it begins growth higher in the water column and earlier in the year, and can therefore establish a dense population that hinders plant growth from seeds or turions (overwintering buds) on the bottom, where light is scarcer. Microalgae can also take up available nutrients rapidly and deny supplies to later-developing plants. Algae also lay down a less structured sediment than plants, smothering seedlings more easily. Importantly, without plant structures, the more effective grazers, the water fleas, are readily taken by the fish, leaving a more inefficient grazer community of rotifers and copepods (Fig. 7.8). The differing nature of the sediment may also lead to greater risk of deoxygenation, which particularly disfavours piscivorous fish, under algal-dominated conditions, and allows proliferation of small, zooplanktivorous fish, thus redoubling the impact of the fish on the grazer community.

In each case the buffer mechanisms will maintain their communities unless something happens to stretch the buffers too far, when the system may abruptly switch to the alternative (Fig. 7.9). The plants might be damaged by being raked out or by deliberately applied or stray herbicide, or by boat propellers; they might be overgrazed by geese

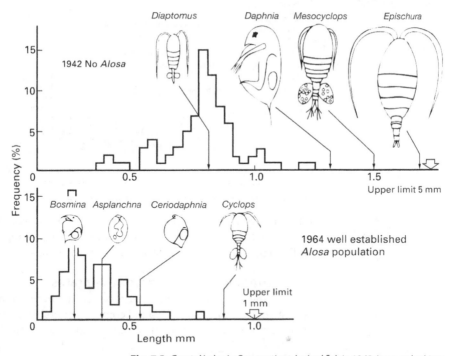

**Fig. 7.8** Crystal Lake, in Connecticut, lacked fish in 1942. Its zooplankton included many large copepods and cladocerans, which had a modal length of 0.8 mm and a maximum of 5 mm. A zooplanktivorous species of fish, *Alosa aestivalis*, was introduced and when the community was examined again in 1964, the mode had fallen to 0.3 mm and large species had disappeared. The largest was a copepod at 1 mm. Modified from Brooks & Dodson (1965).

and swans, perhaps built up into excessive populations by feeding with bread or cake (commonly on the ponds of municipal parks), or by potatoes or other food used to boost their numbers as a public spectacle on some bird reserves; or they may be uprooted by common carp. Pesticide runoff or a myriad of organic substances and other toxins that are present in farm and domestic effluents may damage the grazers on algae.

All of these are switches dependent on the sorts of human activities discussed in Chapter 8, but there are some natural ones too. Changing weather from year to year might result in flooding and increased water levels that disfavour the plants by reducing light at the bottom, or by drought that reduces the number of surviving plant propagules in the sediments. A still summer night may lead to enough deoxygenation to kill piscivores and allow zooplanktivorous fish populations to increase. Alternatively a fall in water level

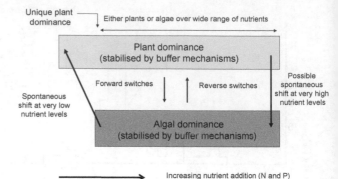

**Fig. 7.9** Alternative states hypothesis in shallow lakes.

short of drying out may favour the plants, or a very severe deoxgenation can remove all the fish in a mass overnight kill, favouring the zooplankton and then the plants.

## 7.4 Experiments with alternative states

One of the first experiments that revealed the importance of fish in determining how a pond community is structured was carried out in artificial ponds: children's inflatable plastic paddling pools, 2 m in diameter and 30 cm deep placed on a flat roof of San Diego State College in California (Hurlbert *et al.* 1972). The pools were filled with tap water to 20 cm, and a 3-cm deep layer of sand, and each given a litre of dry alfalfa pellets to provide nutrients. To each was added a small volume of plankton concentrated by a net from a nearby pond. Then to each of three of the ponds, fifty *Gambusia affinis*, a tiny omnivorous fish, were added, and three were left without fish as controls. Over a couple of months the two systems diverged greatly from each other. Those with fish acquired a dense microalgal community, a zooplankton largely of rotifers and a light growth of filamentous algae. Those without fish had much clearer water, very small microalgal populations, dense growths of filamentous algae and a community of bottom inverte-brates including oligochaete worms, flies and mayflies. A few stoneworts germinated and managed to survive in all the ponds, because the water depth was so low.

Variations on this experiment are legion and easily ac-complished in buckets or bowls. It may even be possible to create deeper and larger ponds by using watertight plastic containers (now sold for diverse domestic uses such as storage or manoeuvring laundry), and even bigger ones such as water barrels or large plastic dustbins, which are readily available. There is still a lot of controversy about how widely the alternative states hypothesis applies. If you can establish a dense plant community with associated animals in the containers, you could try increasing the nutrient levels (adding P or N separately or together) to find out if the plant community will be able to resist very high nutrient levels without disappearing. There is evidence that increasing nitrate reduces diversity of plant communities and if you can establish a mixed community, you could test this. Floating plants like duckweeds also appear to be favoured by high nutrients and more needs to be known about competition between duckweeds and submerged plants. Always remember to replicate your treatments, however. A minimum of six ponds is needed and ten is better so that you can have three to five each of controls and treatments.

Then there are potential experiments with fish. The best experiments take note of natural fish population densities and you must not subject the fish to conditions that they would not otherwise experience in the wild. In the San Diego experiments the fish densities may have been abnormally high, though very large densities of tiny fish that reproduce very rapidly are common in warm lakes. Indeed they may be so abundant that zooplankton has little chance of surviving and many sub-tropical ponds have dense algal communities or floating plants and few submerged plants as a result. In the UK, the most useful fish for pond experiments are goldfish (though they are not native) or three-spined sticklebacks (*Gasterosteus aculeatus*) (Fig. 3.5), which are native. They are very tough fish and breed easily, so that you can create a supply culture without continual raiding of local ponds. In experiments we usually use all-male populations so that there are no complications of changing numbers through breeding, and inevitably some fish die and need to be replaced. Male sticklebacks are more brightly coloured than the females. Natural fish communities have several species, of course, but this is difficult to mimic in experimental ponds. They are simply not large enough, but, as in all experiments, reality has to be sacrificed to some extent for the sake of control. Natural densities of fish are generally around 10–40 g per square metre, which translates to about one to four small fish per square metre. In bucket ponds, one fish is sufficient. Sometimes changes in plant growth can upset experiments, so you may wish to use plastic plant substitutes. Weighted strips of nylon netting or weighted polypropylene rope, dangled from bamboo canes placed across the rim are convenient. They won't disappear of course!

There are further dimensions to this idea of alternative states. The probability of flipping from a plant-dominated state to an algal-dominated one seems to increase as the nutrient levels increase and the probability of moving back to clear water from a turbid state as a result of fish removal (a process called biomanipulation in full-scale lake restoration) increases if nutrients are reduced. If the nutrients are reduced to very low levels, the pond or lake appears to be able to move back to a clearwater state without any other intervention. The controversial issue is at the other end and one view is that nutrients alone can shift a clear water state with plants to a turbid state, without any of the switch mechanisms discussed above. The problem is that where nutrients increase in real ponds and lakes, they come from agricultural runoff, wastewater or stock effluent and none of these is a pure nutrient solution. Runoff contains traces of pesticides, and effluent a large variety of organic chemicals, some of them hormones and pharmaceuticals. *Daphnia* and other water fleas are easily poisoned by small quantities of these, so it has been impossible to test this idea critically using existing sites. Meanwhile the idea has spawned a great interest in the theory of alternative states, with the idea that there is a hysteresis (see Fig. 7.10) in the concentrations of nutrients that first of all cause the switch to turbidity and those needed to reverse it. A much lower absolute concentration is supposedly needed to move back to clear water than was needed to move to turbid water. Experiments to test this would need access to laboratories capable of carrying out nutrient analyses. Apparent examples of this hysteresis are on record (Fig. 7.10), but we cannot be certain that they are not influenced by external switch mechanisms.

**hysteresis**
the phenomenon of a property lagging behind changes in the factor causing it (e.g. stonewort cover lagging behind changes in phosphorus level)

Nonetheless, this controversy about nutrients has not prevented the proposal of a great deal of additional theory. The switch, when it occurs, is often (though not always) abrupt, occurring over a period of only a few months or a year or two, so that the switch is sometimes called a tipping point. The operation of the buffer mechanisms prior to a tipping point is referred to as resilience and the ideas are being applied to other complex systems such as the alternatives possible in the organisation of human societies. Is our present sort of developed society, for example, prone to collapse and will this be inevitable owing to internally generated mechanisms (epidemics, corruption, war, for example) or will it be driven by external switches such as climate change? Those possibilities have in turn driven research into whether there are symptoms in natural systems

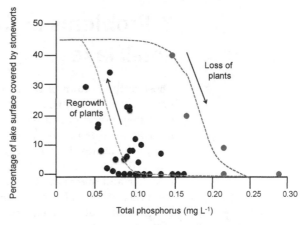

**Fig. 7.10** In Lake Veluwe, in the Netherlands, progressive pollution by nutrients was associated with loss of charophytes (some flowering plants persisted). The phosphorus concentration at which the charophytes disappeared was somewhat greater than the reduced concentrations that had to be achieved to restore the charophytes. This is known as a hysteresis effect, after a similar phenomenon in demagnetising and remagnetising an iron bar. Based on Scheffer *et al.* (2001) from data of M.L. Meier.

that give early warnings of an approach to a tipping point.

Much of this research has been carried out on very simple laboratory systems, computer models and very sophisticated analysis of long-term data sets, and it is still very controversial. For example, if a culture of a cyanobacterium is stressed by being given a short flash of intense light, it stops growth, but then recovers. But if the process is repeated, the recovery is progressively slower and eventually the culture will not recover. It has passed a tipping point predicted by a progressive reduced ability to recover from small shocks. Other research has shown that in approaching a tipping point there is much greater variability in aspects of the system. A fall is preceded by an increasing tendency to wobble. Thus in a plant-dominated system, the algal populations in the water, normally kept low by the buffer mechanisms, may show increasingly more frequent spikes of short-term increased populations. In social systems these symptoms might be reflected in outbreaks of new diseases or financial crises and progressive difficulties in controlling and solving them. This begins to depart far from an interest in ponds, but it illustrates how such interests can lead to much bigger ideas. Concerns over the loss of plants from small lakes and ponds were pivotal in the development of general ideas about resilience, tipping points and our future.

# 8 Problems with ponds and small lakes

Those at the bottom of the heap suffer most. That could not be truer of freshwaters, for they lie at the bottoms of catchment areas and suffer the consequences of any mismanagement or misuse of the land surrounding them, and of the atmosphere from which rain falls, directly or indirectly into them. Landscapes are never unchanging, however, and there have been times in Earth's history that were very different from those now. There have been great extremes of heat and cold, and coincident with them and with changes between these extremes, have been major changes in freshwater habitats and the organisms that live in them.

In Britain and Ireland the scene was more or less completely reset by the glaciation that began finally to dwindle, following many retreats and advances of the ice over nearly three million years, about fifteen thousand years ago. The ice in the last half a million years or so had not completely covered Britain. Large areas of southern England had been unexposed though they had been frozen for much of the year with brief thaws in summer (Fig. 8.1). There had been liquid water in a landscape that resembled the present tundra regions and there would have been ecosystems in rivers of melt water, floodplains, pools and ponds, and even some large lakes ponded up against the ice front. It is likely that many microorganisms persisted from previous interglacial warm periods in these areas and also many invertebrates and even some fish, like the whitefish now confined to a very few of our larger and cooler lakes. However we have few sedimentary deposits, with their pal-

**Fig. 8.1** Ice Age landscape. Painting by Mauricio Anton.

aeolimnological evidence, left from these periods because the last major melting back of the ice covered the land with gravels, sands and clays washed out from under the glaciers and eroded them away. Our best evidence comes from studies of the nucleic acids of our particular races of some organisms, their differences from mainland continental stock and knowledge of evolutionary rates that depend on changes in DNA. These suggest that bullheads (*Cottus gobio*), for example, managed to persist in the UK during the glaciation.

The ice had tied up a great deal of water, so sea levels were much lower than at present and Britain and Ireland were connected by land to what is now the mainland continent. Little is known of the connection with Ireland for it was severed by flooding early after the ice began finally to melt, but explorations in the North Sea have allowed some reconstruction of the plain connecting Britain with The Netherlands and Germany, across which English rivers flowed as tributaries of the Rhine. To the west, migratory fish like Atlantic salmon (*Salmo salar*) and eels (*Anguilla anguilla*) gained greater access and to the east, other organisms were able to move up the tributaries and recolonise Britain, a process that was interrupted about 8,000 years ago when ice melt and rising sea level covered the Doggerland Plain. This formed the North Sea as a connection between the North Atlantic and the English Channel, formerly just an inlet of the Atlantic.

It was still cold but very wet, even as temperatures increased and forest began to cover the land. The lowland rivers naturally swelled in winter and spring, the water being accommodated in wide floodplains. Shifting of deposited silt, variations in current flow, and establishment of wetland vegetation ensured an ever changing system of meanders and cut-off oxbow lakes, pools ponded back by ridges of gravel and sand, and wet grasslands towards the landward edges. North-west Europe is naturally wet, with high rainfall from the Atlantic winds and a northerly latitude that keeps temperatures, and therefore potential evaporation, moderate in summer. Pools and ponds abounded in every depression, especially on the less porous rocks. It was a very wet land, with strong connections between rivers and streams, wetlands, lakes and ponds and their communities. Salmon and sea trout freely moved up most of the rivers to spawn in the headwaters. Eels must have been exceptionally common. The isolation of the mainland from the islands stopped a major route for immigration and as a result the biodiversity of Britain, and even more that of Ireland, is lower than that of the continent, both on land and in freshwaters. Invasions of birds and winged insects continued to be possible. Large herds of large mammals roamed the land, transferring nutrients in their dung, and swarms of midges on their mating flights fed birds, bats and spiders for up to several kilometres from the water's edge.

The developing deciduous forests, probably more or less continuous, but interrupted by areas of grassland maintained by the grazers, or too wet for trees, provided huge amounts of organic matter as windblown leaves to the streams. Blown-down trees covered many reaches of streams and small rivers with woody debris, some of it to be washed downstream to form temporary dams that ponded back water but which were easily burst in the next flood and no barrier to the migrations of fish. Likewise beavers blocked the flow with their dams, creating a more varied habitat but not forming any permanent disruption to fish movements. The waters were at first mineral and nutrient rich from the availability of fresh rock debris for leaching and from the redistribution of nutrients from the land by the grazers. Only the mountain tops and steep slopes and screes were bare of trees but as the weather warmed and rainfall increased about 3,000 years ago, the flatter upland plateaux began to develop blanket bogs that were extremely wet and may have displaced trees from these areas, though possibly there were human influences in that transition.

People moved back as the ice melted. The periglacial landscape at the height of the ice advance would have permitted summer forays but probably these were not really worthwhile. When the ice began to melt back finally and the summer became longer, men and women followed animal herds as they hunted and gathered from temporary camps in caves and soon established more permanent villages, from which they carried out their activities in what we now call the Mesolithic phase. They had boats; there would have been movements of new bands; news and immigrants would arrive, bringing the innovations of stock-keeping and eventually crop husbandry that had developed far to the south in the Mediterranean region. Dogs, domesticated from wolves, had been around for some time. Other stock, derived from wild horses, aurochs, wild boar and ducks was adopted and domesticated. We do not know how, but orphaned young, kept as pets might have been their origins, or clearings in the forest may have been made to attract the less wary of the grazers closer to the camps. Hunting and gathering gradually gave way to more settled existences, and eventually to the penning of stock and the tillage of land for a more convenient supply of grain and vegetables.

Somehow the lesson that animal manures placed on the fields led to maintenance and increase in crop yield was learned. The weather had warmed considerably by the Neolithic, when serious agriculture had been established, along with more ordered settlements of stone, elaborate graves for the great and the good, and stone circles for ceremonies. These sites were not randomly distributed. Many were on the better-drained areas where the woodlands were lighter and the soils more easily workable. They were on the chalk lowlands and sandy areas and on the more moderate, well drained slopes of the uplands or the flat plateaux. Many also took advantage of the defensive advantages of floodplains where platforms could be built above the expected water levels and where the rich vegetation attracted wild game and provided fertile grasslands for stock (Fig. 8.2). The heavy soils of the centre of England, with their dense oak woodland, were avoided. Wooden or stone ploughs could not make inroads into the soils and hardwood trees defeated blunt stone and soft bronze axes. It was not until about 3,000 years ago and the

**Fig. 8.2** In the Somerset Levels, near Glastonbury, lies what is now an unprepossessing drained field (right). But underneath it are the remains, excavated by Arthur Bulley in the early twentieth century, of a vigorous village (reconstructed on the left), dating back to 250 to 50 BCE in the Iron Age, where in the floodplain of the River Brue, before the area was drained for agriculture, the local game and some tended stock provided the needs of some 200 people.

introduction of iron tools that these areas could be cleared. By that time also, the weather had become even wetter and there was a prerogative to drain many areas for cultivation to continue. It was the start of a conversion of a wetland Britain to something much, much drier.

Drainage, or attempts at it, began at least before the Roman Period. The Romans built a long wall bordering the Wash to prevent marine flooding and allow salt marsh to become arable after the rain had leached the soils. The Saxons built low embankments along rivers, again with the intention of preventing flooding. The Danes dug peat from the floodplains of the Broadland rivers during a relatively dry period. When the weather worsened in the thirteenth century, the pits were flooded and connected to the rivers, and became a series of shallow lakes, the Broads. But the trend was towards drainage rather than creation of new waters. Drainage was very much on a minor scale until the invention of steam pumps and the harnessing of coal in the eighteenth century allowed widespread major attempts to drain floodplains. There was embankment and deepening of the rivers and loss of the myriad lagoons, ponds and lakes that had populated the floodplains. Some water meadows persisted because of the beneficial effects of silt from water that flooded over them in winter, but few now remain and there are no British floodplains whose water levels are not controlled in some way.

On the upland plateaux the demands of sheep farming and grouse shooting have also led to widespread drainage and drying out of the peat so that in most areas the growth of *Sphagnum* and accumulation of peat have ceased. On the positive side, however, in the eighteenth and nineteenth centuries there was widespread creation of small artificial ponds as listed in Chapter 1. Even if the river flows had been tamed by early drainage and river engineering, water was still needed for every rural and urban industry. Conurbations, and the epidemics they fostered through poor sewage disposal and contamination of drinking water supplies, led the Victorians in particular to build many dams and create small reservoir lakes in the uplands, but by the early twentieth century water was not nearly so abundant on the land as it had been a thousand years earlier and worse was to come.

The Second World War led to food shortages in Britain. After it there were pressures for much more intensive farming, greater fertilisation and more drainage. Small farms became unviable as world markets drove prices down and grain crops could be grown more profitably on specialist farms with investment in large machinery. Many ponds were filled in so as not to hinder the passage of the machines, and tile drainage led to falls in water tables so that those that survived more often dried out. Small rural industries with ponds for uses like flax retting and steam hammers were no longer needed as industries were consolidated into bigger factories. Stock farms began to rely more on water piped from the mains to drinking troughs; ice was produced in refrigerators not ice ponds, and even in the north of Britain, blanket bogs with their many small pools were drained for forestry that was profitable only because of very high subsidies, and grouse culture that serves only a rich minority. The map of Britain had changed from one with extensive areas of blue for freshwater habitats to one that was predominantly brown.

There have been some reverses. Gravel workings in the floodplains eventually are abandoned when local demand abates or good deposits are worked out and many naturally then flood from ground water. Fashions in gardening have led to creation of many ponds but very small and highly controlled ones, with the water held in by liners and the plants grown in pots lest they get out of control. A tendency to cover gardens with paving, the spread of

**Fig. 8.3** Britain now has a much drier landscape than it would naturally have. The water table is generally lowered and carbon storage is inhibited. Paving drains water directly to the sea rather than into the ground; watering of lush golf courses loses much water to evaporation in summer and draining of moorlands and encouragement of heather by burning for grouse culture removes water storage and encourages erosion of the drying peats.

concreted development and a proliferation of golf courses has hindered recharge of the groundwater and reduced the water table in many areas. By and large the net trend in wetness of the land is still downwards (Fig. 8.3).

The quality of what remains is also much less high than it would have been a thousand years ago, though in the cases of some rivers and canals that were desperately polluted by raw sewage, much better than it was 100 to 150 years ago. Alas, in water quality problems we have solved the nineteenth century problems only to replace them with new and less tractable ones. Changes in water quality began as soon as the postglacial period began to unfold, but have to be seen against our current preferences. Early waters were richer in nutrients than they were to become but we have adopted the idea that high nutrient levels are always bad. Sometimes the concentrations are now excessive and are problematic, but the scale is not absolute. In some areas, with large herds of grazers on naturally fertile soft rocks or glacial drift, the nutrient levels would have supported eutrophic, possibly even hypertrophic conditions, with all the characteristics we would now deprecate.

We do not know exactly what conditions would pertain had we not invaded and changed every landscape on Earth. Even remote Antarctica shows evidence of human activity in traces of airborne pesticides. Antarctic fur seals, once nearly hunted out, have increased in numbers as whaling has relieved competition for food and the seals haul out near coastal lakes that they eutrophicate with their droppings. But the general trend has been a steady increase in nutrients from the time that farming first developed because removal of natural vegetation destroys also the mechanisms that retain nutrients within the soils and biomass of the land. When natural forest is felled, there is an immediate pulse of nutrients lost to the streams and the soils very soon become so depleted that crops cannot be grown on them after a very few years. That is the reason for shifting cultivation in tropical forests where cleared patches are quickly abandoned and new ones created whilst the old ones recover (over a much longer period). Permanent agriculture means continual fertilisation; simple crop systems do not have the mechanisms evolved in natural, complex vegetation to retain and recycle nutrients.

Where land was converted for villages and towns, there are two processes that increase nutrients in the local freshwaters. First, water tends to run off compacted and then paved surfaces, leaving no opportunity for absorption of some nutrients back into the soil and secondly, food is brought into settled areas form a wider hinterland. Nutrients in the food are effectively concentrated from a wider area into a smaller one, but then excreted locally, giving high concentrations both in raw sewage and food wastes and treated wastewater, even when advanced processes have been used to remove a high proportion of the nutrients as a result of operation of Directives such as that for European wastewater treatment.

## 8.1 Creating your own nutrient studies

Nutrient pollution (eutrophication) is associated with decreased biodiversity because competitive plant and algal species can take advantage of the high supply and spread at the expense of slower growing, less aggressive species. Within either of the plant-dominated or algal-dominated states discussed in Chapter 7, there is a trend of reduced diversity as nutrients increase. You can investigate nutrient distributions without having the means to analyse nutrient levels by using desk studies or proxies: other variables linked with nutrient availability. Desk studies mean exploiting the large amounts of information now available on the web sites from the environment agencies (Environment Agency in England, Natural Resources Wales, Scottish Environment Protection Agency, Northern Ireland Environment Agency, Environment Protection Agency, Ireland and the European Environment Agency). The European Agency is much less coy about making detailed information available and has excellent maps showing nutrient levels that you can relate to knowledge of topography, population density, and climate. The national agencies also have such maps but they require more searching. You can use Freedom of Information requests to obtain more detailed data.

To look at the links between nutrients and biodiversity, try determining the catchment area of a set of small water bodies from large scale Ordnance Survey maps. This can be done by inspecting the pattern of streams and the distribution of the contours. Choose a catchment that is neither very small nor very large, say one of less than 50 kilometres square. Then, in the field, estimate from vantage points in the catchment the proportion of land in crop agriculture, pasturage, and semi-natural vegetation and make a count of numbers of stock that you can see, and of houses. Assume each house contains three people on average. Table 8.1 then gives working figures for how much nutrient will be running off from different land uses and from different

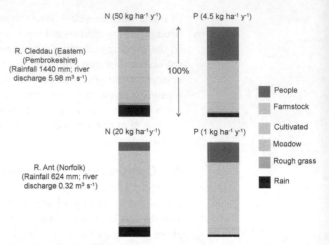

N (50 kg ha⁻¹ y⁻¹)    P (4.5 kg ha⁻¹ y⁻¹)

R. Cleddau (Eastern)
(Pembrokeshire)
(Rainfall 1440 mm; river
discharge 5.98 m³ s⁻¹)

100%

People
Farmstock
Cultivated
Meadow
Rough grass
Rain

N (20 kg ha⁻¹ y⁻¹)    P (1 kg ha⁻¹ y⁻¹)

R. Ant (Norfolk)
(Rainfall 624 mm; river
discharge 0.32 m³ s⁻¹)

**Fig. 8.4** Nutrient sources in two British catchments. A contrast is shown here between a very wet, hilly catchment of the Eastern Cleddau in Pembrokeshire and the very flat River Ant catchment in Norfolk. The total amount of nutrient running off each hectare is greater in the Cleddau because of the slopes and higher rainfall, but the concentrations in the water will be lower because of the much greater rainfall and greater dilution. Stock and people dominate both the phosphorus and nitrogen loads, whilst cultivated land is important for nitrogen in the more arable catchment of the Ant. Based on Johnes *et al.* (1996).

sorts of stock, from people and entering in rain. Calculate the total load for the whole catchment. From meteorological data for the area, you can refine this to calculate likely concentrations by dividing amounts running off the catchment (the total load) into streams and ponds by the difference between precipitation and evaporation (the runoff) multiplied by the area of catchment. Be careful about units. It is easy to make mistakes when moving between square kilometres, hectares, metres, kilograms and milligrams. You can then relate this information to your observations of numbers of species of plants, algae or invertebrates in standardised samples from the ponds and can compare different catchments. This will also give insights into the main sources of nutrients for different sorts of catchments. People will dominate in urban areas; often it will be farm stock in northern and western Britain, or arable cultivation in the south and east (Fig. 8.4).

Eutrophication from agriculture gradually increased as more and more fertiliser was used from the nineteenth century on. Prior to import of nitre from guano islands off South America, the only fertiliser available was from recycling of stock wastes and farms had to be mixed for that purpose. Their overall fertility tended to run down with time because nutrients were continually lost to the streams and could not be replaced rapidly enough by rock weathering and nitrogen fixation, though the growing of legumes helped. In the late nineteenth and early twentieth centuries the development of superphosphate by adding sulphuric acid to phosphate rock, and the Haber process for producing ammonia, offered cheap ways of bringing in large quantities of fertiliser and intensified crop growing and the run off of waste nutrients to freshwaters. The expansion in population, which this increased production could support, meant a greater flow of excreted nutrients to the system also. Currently the levels of total nitrogen and total phosphorus (which include all forms, inorganic, organic and particulate present in the water) in agricultural and settled catchments in Britain and Ireland are generally 10 and up to 100 times those we might expect from catchments with their natural vegetation intact. For those catchments we have to look to other countries like the USA, where there are still large tracts of near wilderness.

## 8.2 Minimising eutrophication

There is a range of measures that landowners can use to decrease the loss of nutrients from their land, but substantial losses remain inevitable once the natural vegetation is destroyed. The obvious first way of minimising losses is to restrict fertiliser use to what is absolutely necessary but no more. Fertiliser is relatively cheap and there is always

**Table 8.1** (p. 174) Data for the calculation of nutrient loads and thence concentrations in runoff in pond catchments. Values are given per hectare per year for rainfall and dust and land uses, and per individual per year for farm stock and people. To calculate the total load remember to think in terms of the whole catchment and to calculate for actual area and for actual numbers of stock and people. Allowance has been made for appropriate breeds and for the topography and vulnerability to erosion in different areas. The total annual load will be in kilograms per catchment. An average concentration in the stream and pond water can then be calculated. First calculate the total runoff (annual precipitation minus annual evaporation (both in metres per year), multiplied by the area of catchment in square metres). This will be in $m^3$ per year. Concentration will then be total annual load divided by total runoff and will be in kg per $m^3$. This equates to grams per litre and is multiplied by 1,000 to give mg per litre or 1,000,000 to give µg per litre, the units usually used.

| Source | Annual phosphorus load (kg per hectare for land use or kg per individual for animals) | | | Annual nitrogen load (kg per hectare for land use or kg per individual for animals) | | |
|---|---|---|---|---|---|---|
| | Upland areas (>200 m) or with extensive farming | Lowland rolling areas (<200 m) or with intensive farming | Very flat lowlands or with industrial farming | Upland areas (>200 m) or with extensive farming | Lowland rolling areas (<200 m) or with intensive farming | Very flat lowlands or with industrial farming |
| Rainfall and dust | 0.2 | 0.2 | 0.2 | 25 | 25 | 25 |
| Cultivated | 0.2 | 0.9 | 0.4 | 10 | 50 | 30 |
| Grassland (temporary and permanent), conifer plantation and orchards, public parks | 0.3 | 0.8 | 0.4 | 2 | 30 | 10 |
| Housing, streets, industry | 0.3 | 0.8 | 0.4 | 1 | 15 | 5 |
| Rough grazing, woodland, other natural/semi-natural vegetation | 0.02 | 0.07 | 0.03 | 1 | 13 | 3 |
| Cattle and horses | 9 | 18 | 12 | 40 | 80 | 60 |
| Pigs | 6 | 6 | 6 | 19 | 19 | 19 |
| Sheep | 1.5 | 1.5 | 1.5 | 10 | 10 | 10 |
| Poultry | 0.2 | 0.7 | 0.4 | 0.6 | 0.6 | 0.6 |
| Persons (urban, served by mains sewerage) | 1 | 1 | 1 | 4 | 4 | 4 |
| Persons (rural, served by septic tanks) | 0.4 | 0.4 | 0.4 | 2 | 2 | 2 |

a tendency to add a little extra. If manure is used, it helps to apply it in dry weather and to plough it in. Manure is also best stored for reuse in covered silos rather than open heaps and the same is true of silage, from which trickles of extremely nutrient-rich water will find their way downhill to the nearest stream in wet weather. Ploughing along the contours of the land rather than up and downhill, even if this means more turns for the tractor, reduces soil erosion and it is in eroded soil that much nutrient comes. Not ploughing at all but direct drilling is better though it may exacerbate other problems when herbicides are used to control weeds.

Relocating gates for stock at the top of fields, so that soil from the trampled areas close to the gates washes into the field rather than out of it, also helps. There is a notion that leaving a strip of land uncultivated parallel to streams stops the runoff of nutrients. It might if it is wide enough (10–20m compared with the 1–2m usually left) but it can be bypassed by water moving underground through soil channels and it works best for nitrate, which can be denitrified if the soils close to the stream are waterlogged. It works much less well for phosphate because the soils become saturated and the waterlogged conditions that favour denitrification favour release of phosphate. By and large, although these approaches may reduce nutrient levels, they have proved less effective than hoped and coexistence of modern agriculture with high biodiversity in freshwaters is impossible to achieve without very wide buffer zones between the two. The ultimate and conceptual problem is that the nutrient levels needed to support high biodiversity are very low and incredible to farmers who would regard much higher levels as deficient for any sort of crop growing.

## 8.3 Acid rain

Eutrophication is probably the most pervasive problem for freshwaters everywhere but not the only one. We are still coping with the problem of acidification in the UK and North America and it is becoming prominent in newly industrialising areas of Asia. It arose with the extensive burning of coal in the middle of the nineteenth century. Coal produces waste sulphur gases that oxidise in the air to sulphuric acid. The acid impacts as droplets on vegetation and snow, or washes down in rain. Ammonia released from coal burning also caused acidification by oxidising to nitric acid. Uncontaminated rainwater has a pH of about 5.3, slightly acid because of the carbon dioxide that naturally dissolves in it. Acid rain may have pH values of 2 or occasionally lower, but generally around 4. Such low pH values cause problems for invertebrates and fish either directly or because the acid water readily mobilises aluminium salts from the soil minerals. Aluminium ions are very toxic, causing respiratory difficulties in fish, and reducing eggshell thickness in water birds feeding on freshwater invertebrates. Low pH eliminates crustaceans and snails, which cannot deposit sufficient calcium to form their carapaces and shells, and prevents fish larvae from being able to break through the egg membranes.

Acid ponds and lakes may be very clear because the aluminium flocculates particles in the water and precipitates phosphate, but flocs of cotton-wool-like filamentous green algae (particularly *Mougeotia, Zygnema* and *Spirogyra*) may become prominent, fish disappear and acid-tolerant species of *Sphagnum* moss replace a more diverse plant community around the edges. As a result of wind drift of acid from the tall chimneys of British industry and power generation, there were widespread losses of fish communities from Norway and Sweden in the late twentieth century. Tall chimneys had been seen as the solution to low-level chimneys and formation of sulphur dioxide smogs at ground level in British cities, which caused widespread bronchitis in their citizens.

Solutions that treat symptoms are often apparently cheap but may be expensive in the long-term. The best solutions are those that treat causes. Eventually international pressure forced such solutions through absorption of sulphur gases, at power stations and combustion plants, into powdered limestone and through use of low sulphur fuels. This had been preceded by attempts to dump lime onto catchments that were particularly vulnerable to acidification because of the nature of their rocks, or into lakes themselves. These were only short-term in effect but often destroyed specialist acidic communities in the uplands. Unfortunately, the acidification problem is not entirely solved. The response of stream and lake communities to curbing of sulphur emissions has been slow, despite marked improvements in water chemistry, primarily reductions in sulphate concentrations. This may be because pockets of acid water remain in the soils and are washed out in heavy rain, but it is also because a new source of acid has taken over from the old. Vehicle exhausts produce nitrogen oxides that oxidise to nitrate, and intensive stock units, such as chicken batteries and mechanised dairy operations, release

**flocculates**
forms flocs (accumulations or clumps of particles)

**precipitates**
produces a solid deposit from a solution

a lot of ammonia that similarly oxidises. The problem is not as serious as it was but rain is still often of lower pH than its natural 5.3 and catchments with poorly buffered soils still suffer. A box of pH papers and a conductivity meter are useful portable tools in the uplands. On casual walks a picture can be built up of local conditions. There is, however, an interesting complication of the behaviour of *Sphagnum* mosses in ponds and pools of the uplands.

## 8.4 The interesting peculiarities of *Sphagnum*

*Sphagnum* moss (Fig. 8.5) is one of the world's most abundant plants. There are about 380 species worldwide and collectively *Sphagnum* plants cover a huge area of the wet tundras and forests of the northern world. They have particular properties that allow them to outcompete other plants, provided water is abundant (they do not tolerate drying). These are that they can adsorb mineral ions onto the sides of their cells and replace them with hydrogen ions. This

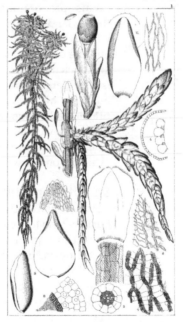

SPHAGNUM AUSTINI.

**Fig. 8.5** *Sphagnum* structure drawn in 1864 by William Starling Sullivant. An entire plant is shown at the top left, with the microscopic structure of the cells at the bottom right. The capsule, in which spores are produced, is in the centre at the top and details of the leaves are variously shown. Sullivant was a surveyor and engineer but became, as an amateur naturalist, the leading American expert on mosses and liverworts of the 19th century.

**adsorb**

accumulate material as a thin film on the surface of a solid

maintains the low pH that prevents many other plants from colonising. If you examine some *Sphagnum* leaves under a microscope, you will see small green cells and very large, now dead cells (Fig. 8.5, bottom right). It is on these dead cells that the ions are adsorbed and which also soak up a lot of water that gives *Sphagnum* its spongy feel. The maintenance of acid conditions also slows decomposition and allows *Sphagnum*-dominated lands to store a great deal of carbon, something that is very important in maintaining the Earth's carbon balance (see next Chapter). If you bring back some fresh *Sphagnum* and pack it into a jam jar and flood it with water that is not acid, perhaps some bottled water from a limestone spring or your tap water (if first you allow it to stand to lose any free chlorine, used in some areas to disinfect it), you should be able to demonstrate a fall in pH of sometimes one or two units compared with a control not containing *Sphagnum*. Sometimes the pH will fall well below that of rain. If you live in an area where *Sphagnum* occurs (predominantly the uplands but also on heaths in the lowlands where the rocks and soils are not rich in lime), it is worth looking at the acidifying properties of *Sphagnum* and the distribution of pH among different species. Some are much more adept at acidification than others. *Sphagnum* is considered something of a challenge for identification, but as in all 'difficult' groups, a little perseverance and practice will soon overcome the problems and can make you a very valuable amateur expert in a world where professional taxonomists are dwindling fast. There is a free key available on the web site of the British Bryological Society.

## 8.5 Other problems

There are many other pollutants that influence freshwaters, but they tend to occur at very low concentrations and require advanced equipment for their detection. They include herbicides and pesticides, industrial organic substances like polyaromatic hydrocarbons (PAH), dioxins, flame-retardants, and pharmaceutical products, as well as heavy metals. Many of these are very toxic and introduced onto the market, or produced as by-products of industrial processes, with only limited testing. Several hundreds of new chemical are developed every year and by-products are often not suspected or ignored until problems become obvious. Once effects are detected it may be some time before governments are forced into action but eventually many toxins have been taken out of use, though some have long half-lives in the environment before eventually they disappear. Mercury and its compounds were once widely used but are now largely banned by international convention. Mercury still causes a problem however because it is indestructible and traces of it can be concentrated into fish. Mercury used in warm regions volatilises, enters the atmosphere and eventually condenses out in cold regions like mountains or the Poles. Under the European Water Framework Directive, member states now have to produce river basin management plans and these are published on the web sites of the environment agencies. With some patience and persistence it is possible to discover quite a lot about water quality in your local area. The plans are updated every six years, the latest ones at the end of 2015.

From the point of view of ponds, however, the key problems are those of eutrophication, acidification, gross organic pollution by excreta from farm manure heaps (which adds deoxygenation to the problems of eutrophication), and obliteration through deliberate filling-in and drainage. Many pond ecologists would consider the latter the most important immediate problem. It seems incredible that not very long ago wet places were considered waste land to be used for such practices as the dumping of city rubbish. Even now, wetlands, often in river floodplains, are seen as sites to be filled with rubble and then built over, a practice that ultimately leads to local problems when flood defences are overtopped because floodplains have been destroyed upstream. Re-creation of ponds is thus the frontline for pond conservation.

Ponds are rich in species and creation of ponds is something that local organisations can do. Ponds are particularly valuable for amphibians (especially the common frog and the three species of native newt) because they often lack fish, which eat the tadpoles. Of more than 4,000 freshwater macroinvertebrates in Britain and Ireland, 60–70% can be found in ponds. Red Data books, a concept created by Peter Scott in 1963 as a way of cataloguing rare and threatened species in different categories of risk, list two-thirds of 300 such British and Irish invertebrates as found in ponds, though they may occur in larger lakes and rivers too. Some, especially those that are particularly vulnerable to fish predation and which require a period of drying out to compete with others, such as the tadpole shrimp (*Triops cancriformis*), are confined to temporary ponds. It is a measure of the extent of pond destruction that tadpole shrimps (Fig. 8.6) now only occur in two places in the UK, yet hitherto have survived in unchanged form for 200 million years.

**Fig. 8.6** The tadpole shrimp, *Triops cancriformis*, which is an animal of ancient origin and confined to temporary pools. The body length, without appendages, can be up to 5 cm long. Photograph by Gerhard Rothender.

The problems of engineering damage to river systems, through deepening, canalisation, replacement of natural banks with concrete, and isolation from the former floodplain, are thorny and involve major conflicts among different sections of society. They will have to be solved, but by major engineering operations, and local opinion will be very important in finding compromises. But local action with minimal fuss is possible with pond creation and it is a popular activity with amenity, conservation and educational groups.

Pond creation, however, is not simply a matter of digging a hole just anywhere. If the hole is to be bigger than 25,000 cubic metres (equivalent to 1.25 ha with a mean depth of 2 m) or if the damming of a water course is involved, there must be permission from the environmental authorities, and the pond will fall under the Reservoirs Act (1975) and must be constructed under the supervision of a chartered civil engineer. You may need planning permission and will need licences from the national environment agency if the water is to be taken from or discharged to a stream or groundwater, if you wish to stock fish, or if it is fed by a river containing migratory salmonid fish. If it is near a road, you may need permission from the Highways Agency. But in general pond creation is a much more modest operation. Of course ownership (or support and permission of the owners) of the land concerned is needed.

Frequently new ponds are dug out in unsuitable places where the water table is too deep and some sort of

**Fig. 8.7** A badly sited pond in Nevada. Farmstock are eroding the edge, which has no emergent plants. Erosion is giving very turbid water, in which little will grow and the bank edge is too steep to look natural. Photograph by Nevada County Conservation District.

impermeable liner of puddled clay, concrete or butyl rubber has to be used to hold the water. Such garden ponds often keep their owners busy with problems of leakage. On a bigger scale, in nature reserves, the pond often does not look or feel 'right' in its landscape and may be too regular in shape, erring in either roundness or squareness. It is very difficult to get a natural look (Fig. 8.7), and needs sensitivity to the lie of the land and direction of prevailing wind. There is a tendency to try to put in all possible features, from islands for bird nesting (artificial islands erode away if not protected with rock or concrete edges, which look odd) to shallow edges for ducklings to walk in and out, and steep banks for nesting holes for kingfishers and martins. The most sensitively constructed ponds have gently sloping edges (less than 1 in 15) for a natural looking littoral area. Excavated spoil can be used to give irregular surrounds with temporary puddles and there should be interconnected swampy areas for animal cover. Groups of ponds of varying size and depth in a naturally waterlogged area give the greatest diversity, when natural colonisation is allowed and the lure of garden centres with their cultivars of water lilies and various exotic species is avoided. Such an approach is least expensive and what grows is what will grow naturally with no need for the meddling management of periodic dredging, vegetation clearance and general tidying up.

On a bigger scale, there are opportunities to create ponds and lakes, often of several hectares, from exhausted gravel workings. About 100 million tonnes per year of gravel are extracted from floodplains in the UK for the construction industry and 1,500 ha of worked out excavations become available each year, 500 ha of which are floodable. There are also about 60,000 ha of exhausted workings, 15,000 ha of them already flooded. This does not compensate for the rate of loss of other ponds but it is a valuable contribution. New gravel workings can become extremely valuable wildlife habitats, particularly when the gravel is removed 'dry' from a pumped pit. If it is extracted from a flooded pit, the water tends to be turbid from disturbed clays and when the pit is abandoned often remains turbid, because aquatic plants cannot colonise because of the poor light climate. Where it is kept dry but later flooded, aquatic plant colonisation and clear water may be achieved. More than 20% of the populations of several duck species in the UK depend on flooded gravel pit lakes. Perhaps the *pièce de résistance*, however, was the conversion by the Wildfowl and Wetlands

**Fig. 8.8** London Wetland Centre. Photograph by Patch99z.

Trust, of the abandoned concrete Barn Elms reservoirs in central London to a pond and mire area now extremely rich in wildlife (Fig. 8.8). It was remarkable because the project had to be completed, under the planning controls imposed, without substantial import of materials or removal of waste from the site, and was very skilfully completed to have an extremely natural look.

## 8.6 Pond management

Conservation groups often meddle with ponds, by cutting surrounding trees, dredging and deepening them, introducing frog spawn (and sometimes diseases of amphibians) and planting irises and other colourful flowers, especially water lilies, which may not even be native. By and large, natural processes can be relied upon to do the jobs of colonisation and establishment of a persistent community far more effectively. A pond once created is usually best left well alone.

There are many common myths about pond management. These include: *the bigger the better* (very small ponds, however, may harbour the more unusual of crustaceans, able to compete with the more robust ones because they can survive drying out); *ponds should not be shaded by trees* (but trees provide fallen leaves, which are important sources of energy for some invertebrates); *ponds need to be dredged to keep them from being choked by vegetation* (plants provide habitat and 'choking' is a human concept not unlinked with the urge to control things that characterises the more demonic of gardeners); *ponds must have oxygenating plants* (not necessarily; plants produce oxygen by day but consume more or less just as much by night or even on dull days; oxygenation is largely provided by diffusion from the atmosphere, helped by wind).

Then there are the notions that: *new ponds need to be planted because natural colonisation is slow* (but it is really quite fast and the advantage of leaving it to natural processes is that what manages to colonise will generally persist; what is planted-in may often die); *water level fluctuations should be minimised* (no they shouldn't; changes in level open up niches for small plants in the exposed mud, and water level changes are normal features of lakes, to which organisms are well adapted); *livestock should be denied access* (it depends on how heavy the trampling is; sometimes plants depend on the bare mud created by poaching of the edges for seed germination. One of only two British sites for adder's tongue spearwort (*Ranunculus ophioglossifolius*) at Badgeworth in Gloucestershire was almost lost when cattle were excluded, ostensibly to protect the plant).

Finally: *there should be an inflow to prevent them becoming stagnant* (some ponds are rain fed, others groundwater fed and still water is not necessarily bad; what is usually meant by stagnant is not stillness but extreme pollution from farm yards and silage heaps; having an inflow from these makes the problem worse); and *ponds are self-contained islands in a sea of dry land* (but absolutely, like all lakes, they are not!).

All that is not to say that occasional management is not useful. Diversity will be increased by dredging a pond completely silted up, or by tree cutting from a pond so overhung that it is densely shaded. It is to say that continual meddling is to be discouraged and that some specialist organisms need silted up and shaded ponds and that over-zealous management will obliterate these conditions.

The key to having diverse and attractive ponds is not to meddle but to ensure that nutrient pollution from the surrounding land is minimised, that there are other ponds in the vicinity for colonists to arrive from, and that there is variety in the surrounding land (trees, varied topography, and as much wet habitat as possible) – basically the same principles for maintenance of any freshwater system of high conservation value. The best places for creation of new ponds are on river floodplains, with contact of some ponds through flooding with the river and isolation of others to keep fish out: the very places where ponds have been most frequently obliterated. A natural landscape has reason and meaning; a human-dominated landscape often does not and it is this lack of appreciation of ecological processes that is hurtling us towards unprecedented global problems. Pond creation is to act locally; the other side of the coin is to think globally.

# 9 Ponds and the future

Climate change, food security, water supply and biodiversity: these are the ways in which ponds are really crucial for human affairs and in that order of importance when we take a global and future look. We must look beyond the local issues of Britain and Ireland, where ponds are sidelined for conservation and amenity, to the swathes of land in the warm temperate region and dry tropics, where water is increasingly a problem, to the wet tropics where protein is scarce, and to the tundra and boreal regions where many millions of ponds are part of a key system that keeps our climate equable, but which are currently under threat.

Biodiversity has been discussed earlier. There have been many arguments about it. Clearly ecosystems can lose species and continue to function, but natural selection, responsible for the ultimate production of a variety of species, is not a forgiving process. It is ruthless in removing inefficiency, reflected in lesser abilities to contribute to the next generation. If systems have a particular diversity, it is therefore likely to mean something, even if we do not yet entirely understand it. Many species are comparative rarities, occurring very locally for a variety of reasons but collectively add up to a high proportion of the total diversity. Sometimes they may be rare because the physical conditions for their existence are simply unusual. Sometimes they may be new invaders, just becoming established, or species on their way out as conditions naturally change, but many persist in small numbers for long periods and there must be an explanation for this. It gets us into interesting intellectual territory concerning how to reconcile the ruthlessness of natural selection, which should eliminate such minor players, with the maintenance of a biosphere in which many different organisms appear to contribute collaboratively to maintenance of equable conditions for life. In a natural cycle of change (development of a floodplain pond then its obliteration by high floods and later reestablishment of pond conditions elsewhere on the floodplain, for example) some species will thrive at particular stages but be reduced to bit parts in others, but functioning of the system requires that they persist throughout.

Ponds, with their relatively high representation of our total freshwater algal, plant and invertebrate numbers in particular, might thus represent insurance communities for the greater diversity of a changing countryside. They contain a fascinating collection of species of high scientific interest and there is comparatively little known in detail about most of the organisms. A list of species, a little knowledge of feeding habits, and data on changes in numbers are beginnings but not ends. There is an argument that the better a society looks after its habitats and biodiversity, the more likely it will be to treat its own members well. In that sense pond biodiversity is a barometer of the health of our society. Pond creation and study is a very healthy activity and there is abundant evidence that contact with nature has positive health benefits for people.

## 9.1 Water supply

Water supply is another way in which the state of our ponds reflects the health of our system. We consume a great deal of water. Of course we do not consume it in the sense of destroying it, but we divert increasing amounts of it to our own uses and often quicken its passage to the ocean, bypassing natural aquatic habitats such as ponds and wetlands. Drainage has changed from being perceived as highly desirable for agriculture and settlement to being a problem of our times for several reasons. Drainage takes many forms. It might be the interception of rivers for water used for irrigation of fields in dry areas where production of thirsty crops like cereals and roots would not otherwise be nearly as productive; it might be the canalisation of rivers and disconnection of their floodplains behind levees so as to create agricultural land from what were water-storing wetlands replete with biodiversity. On many landscapes, it is the wet areas that focus the diversity, both of plants and animals with their continual need, in most cases for water for drinking. There is not a single large area of substantial wildlife and conservation interest that does not have a functioning floodplain with wetlands and ponds at its heart. Drainage also takes subtle forms. Lowering of water tables by creating hard and impermeable surfaces: roads, pavements, concreted and flag-stoned gardens, all divert water to the drains and sewers and quickly out to sea rather than into the ground. Falling water tables leave pond bottoms high and dry. Removal by pumping of groundwater, at rates that are in excess of rates of recharge, also lowers water tables. Many licences for groundwater abstraction for irrigation allow excessive removal. The lushness of golf course greens is often at the expense of the ponds in the slacks of dunelands. Conservation of ponds is symptomatic of wise use of water at a time when demands are becoming excessive. Power showers, private swimming pools, excessively watered lawns and car washing do not help in a situation where our individual consumption has moved from about 30 m³ per year in the mid nineteenth century to 320 m³ per year now. Even in a relatively wet north-west Europe, there are substantial lowland areas (in the south east of England for example), where water is short.

One of the features of the archaeology of arid areas has been the uncovering of sophisticated systems for managing water: canals, cisterns and storage reservoirs. Many of these fell into disuse with the ravages of war and migration of peoples, but the need for such understanding of how to conserve water is increasingly urgent. Water supply has often been used as an agent for political control of the less privileged of society; drought undermines civilisation. But creation of ponds and small lakes, used multiply for water and food supply, can greatly improve the lot of poor people.

## 9.2 Food security

It is in the tropics and the boreal zone where ponds reach the zenith of their importance for protein supply and carbon storage respectively. The world's wild fisheries are becoming depleted as fishery technology becomes more efficient, there are more ships and expanding markets. Wild fisheries in freshwaters still contribute a great deal because freshwaters tend to be more productive, because of greater nutrient availability, than the deep ocean, but even they have become overfished. As a result, pond culture is now a major source providing about 42% of the total catch, with about two-thirds of the ponds freshwater and one third marine. Human consumption of fish is annually now around 16 kg per person and about 70 million tonnes of fish are produced in ponds. A little less than a third of this is grown without supplementary feeding, so effectively in 'natural' pond ecosystems.

Ponds collectively are mostly used for edible finfish (some 354 species), crustaceans (59 species), molluscs (102 species), amphibians and freshwater turtles (6 species) and edible plants like water chestnuts, lotus and seaweeds, with a minor growing of algae used for nutraceuticals, including *Spirulina*, a cyanobacterium and *Dunaliella*, which is a source of β-carotene for dietary supplements or food colouring. Asia is the major producer with about 88% of total production. Indeed in China it is rare to find any small body of water in which someone is not growing something edible (Fig. 9.1). Europe, particularly Eastern Europe is the next most

**finfish**
fish with fins, as opposed to shellfish or other aquatic animals

**nutraceutical**
products marketed to improve health or prevent disease and usually derived from living organisms, as opposed to pharma-ceuticals which are biologically-derived or synthetic chemicals targeted at cure of specific symptoms or diseases

**Fig. 9.1** In China, almost every pond is used for growing something to eat.

productive but with only with 4.3%, then Latin America with 3.85% and Africa with 2.2%. Finfish are the most important products and yields can be ten to a thousand times higher than in natural wild fisheries. Expectation is that, in the next five years, aquaculture will supply more fish than wild fisheries, virtually all of which are now fully exploited. This is in interesting contrast to predictions made, in 1961, by a very experienced fisheries ecologist, C.F. Hickling, that, given the huge extent of the seas, fish culture would be an important but very minor contributor.

Fish culture ponds were developed in ancient Egypt and China from at least 500 BCE, but in Europe developed from mediaeval stew ponds in which fish caught from the wild were stored to supply protein in winter to monastic and other communities (Fig. 1.7). Later there developed deliberately managed systems often well integrated into village farming communities. In Europe, fish feed only a little below about 8 °C, so there are fewer than 200 growing days per year, but in warmer climates the entire year is available and production can be very high. Nonetheless, central and eastern Europe have extensive pond culture, often using the introduced common carp (*Cyprinus carpio*), but also native cyprinid fish like bream (*Abramis brama*) and perch (*Perca fluviatilis*). In the UK most fish culture is for carnivorous fish like trout, which must be fed protein food and in which there is a net loss of usable protein in producing fish of upgraded quality for a wealthy market.

Elsewhere, use of herbivorous and detritivorous fish leads to net production of high-quality protein. For efficient conversion of available food, a mixture of deoxygenation-tolerant carp species is often stocked, with grass carp

(*Ctenopharyngodon idella*) feeding on aquatic plants and vegetable waste thrown in, silver carp (*Hypophthalamichthys molitrix*) on phytoplankton, common carp on bottom invertebrates and other species, catla (*Gibelion catla*), rohu (*Labeo rohita*) and mrigal (*Cirrhinus mrigala*) on zooplankton. Africa and South America have contributed tilapias (*Oreochromis, Tilapia, Sarotherodon*), which have a better flavour than carps, and Asia, crucian carp (*Carassius carrasius*) and white amur bream (*Parabramis pekinensis*). Only two fish species have been truly domesticated, meaning that breeding is subject to human control. As a result, their behaviour and appearance differ greatly from those of wild stock, and survival of at least some varieties would no longer be possible in the wild. The two species are common carp and crucian carp (golden carp, goldfish), though a few others are virtually domesticated. Wild-type common carp is now virtually extinct in its native range in eastern Europe and western Asia. Its domestication began with the Romans two thousand years ago and it was spread rapidly in mediaeval times and is now ubiquitous. Goldfish were bred in China largely for ornamental purposes.

To boost production, piggeries, and even human latrines, are erected over parts of the ponds to provide continual fertilisation. Raw sewage is no problem if diluted and the ponds are stocked with air-breathing fish like channel catfish (*Ictalurus punctatus*) and murrel (*Channa striata*) originating from deoxygenated swamp habitats. Manure is however better used in improving soil fertility, and inorganic fertilisers can substitute where nutrients are concerned. Farm wastes like oilcake, copra, groundnuts, rice bran and cereal chaff can all be used as supplementary fish food. Microorganisms will readily condition them to forms nutritious to the fish.

There are problems with the transmission of certain fluke and tapeworm diseases but these are easily avoided by cooking. Bigger problems, especially in the Far East, are those of conservatism, lack of capital and security of land tenure, which may all contrive to discourage large-scale pond farming, but it is widely used and can be carried out in heavily polluted water, such as that drained from irrigation areas that may have become quite salty. Cyanobacterial blooms, derided elsewhere, are welcomed in fish culture ponds for their nitrogen fixing properties. Fish culture is particularly popular in ponds in cleared mangrove swamps, though mangrove provides more important services including coastal protection and nurseries for wild fish,

**Fig. 9.2** La Dombe, in south-eastern France, is an area of old ponds, periodically drained for grain cultivation and flooded for fish culture. Photograph by Tourisme France.

and clearance for ponds should be discouraged, though it is perhaps preferable to removal for marinas and luxury tourist developments.

In mainland Europe, pond culture has a long tradition. Freshwater fish of a greater variety of species than favoured in the UK are eaten, and common carp is a particular delicacy. A long tradition of management of water levels, draining down and fertilisation is exploited. There is considerable understanding of how the ecosystem works and, in the 1960s, observations from ponds in the then Czechoslovakia revolutionised our thinking on how zooplankton communities are structured by fish predation. Particularly interesting is the system of La Dombe (Fig. 9.2), in south-eastern France, where there is an ancient series of ponds that can be flooded and dried out by a series of sluices. The water regime is controlled by local custom and law and allows a system of carp cultivation and maize growing to exist. The maize stubble, after flooding and fish-stocking, provides initial food for the carp, and as the stocking rates are not excessive, an aquatic plant community can develop. The carp turn over the soil and mobilise nutrients so that when they are big enough to harvest and the pond is drained, the maize benefits from an initial flush of nutrients, though some fertiliser is eventually needed.

Tilapia are up-and-coming species. They taste good, have no fine intramuscular bones, breed early and easily, thrive on cheap plant and algal food and hence produce high yields. They are tolerant of wide temperature and salinity ranges, are relatively free from parasites and diseases and hybridize readily. This latter helps breeding programmes. Suitable tilapia species include *Oreochromis andersonii*, *O. macrochir*, *O. niloticus* and *O. aureus*. Because they breed frequently, large populations of small fish may result but his can be avoided by stocking predators like murrel (*Channa striata*) with them, and by culturing first generation hybrids which are almost entirely male and do not breed further. Males also grow faster than females. Crosses between female *O. nilotica* and male *O. aureus* produce more than 85% males and if *O. urolepis hornorum* is used as the male parent, 100% males can be almost guaranteed in the first generation. Treatment of the fry with androgenic steroids can also turn all the fish into males but this increases the cost. Fish pond culture is increasingly becoming a technical rather than a traditional operation.

The major reservation about aquaculture is that it almost always uses alien species, not native ones, though frequently native species are just as productive. In every case of introductions, fish have escaped from farms and established in natural waterways, often suppressing the native species. Although there are codes of practice to try to avoid this, they have everywhere been breached. Escape of the bighead and grass carp to the Donghu Lake in Wuhan, China, has displaced some 60 native species and use of the North American crayfish (*Pacificastus leniusculus*) in Europe has greatly reduced numbers of the native European crayfish (*Astacus astacus*) through co-introduction of a fungal disease (*Aphanomyces astaci*), to which it is tolerant and the European native not. In Australia and New Zealand, introductions of alien salmonids have severely threatened the future of the entire group of galaxioid fishes that are endemic to the region.

## 9.3 Climate change

The final relevance of ponds in human affairs is perhaps the most important, the most unexpected, and the most ignored. Ponds are important parts of a remote wetland that covers much of the northern tundra and boreal forests in Canada, Scandinavia and Russia and they play an important role in passively concentrating and storing carbon as organic matter. Debris from vegetation is washed into the streams and eventually rests in millions of ponds and small lakes, where waterlogging, as in the peaty surrounding soils, inhibits decomposition and the carbon is stored. On a world scale we think that every year about 1.9 billion tonnes of carbon in organic matter is washed into freshwaters from the land, 0.8 is respired, 0.9 is exported to the ocean and 0.2

is stored, and the northern ponds are major areas in which these transformations take place.

Their importance is in the balance of planetary photosynthesis and respiration by which the composition of the atmosphere is maintained. Multicellular organisms are much less tolerant of differences in atmospheric composition than microorganisms. The latter were able to thrive in hot anaerobic seas early in the Earth's history, but multicellular organisms, and particularly the larger ones, need high concentrations of oxygen. Most also need a narrower band of temperature in which to thrive and this is maintained by modest levels of greenhouse gases, primarily water, carbon dioxide and methane. Were photosynthesis to be balanced exactly by respiration, conditions could be kept constant, but there are interfering factors. On the one hand, volcanoes and rock weathering release carbon dioxide and on the other a rise in oxygen concentration would increase the risk of vegetation fires, a phenomenon that has had huge effects in past geological time. Our natural systems have developed mechanisms that inadvertently maintain moderate oxygen and carbon dioxide levels and an equable climate.

They do it through a slight excess of photosynthesis over respiration, which keeps oxygen concentrations up, and through biological production of gases like methane, which reacts with oxygen in the atmosphere and tempers oxygen concentrations downwards. Storage of carbon in peats, soil organic matter and lake and pond sediments is responsible for the slight excess of photosynthesis. It keeps carbon dioxide concentrations down and compensates for release of new carbon dioxide from volcanic vents and rock weathering. Our current problem is that we have been burning huge quantities of these carbon stores (coal, oil and gas), whilst simultaneously draining waterlogged land, which constitutes the current main store. Drainage releases the carbon through aeration and oxidation of the sediments and peats.

At present we release about 8.4 billion tonnes (gigatonnes) of carbon into the atmosphere through fossil fuel burning and cement manufacture, and we destroy enough tropical forest each year to release a further gigatonne. Some carbon, about 2.5 gigatonnes, is precipitated to the bottom of the ocean as calcium carbonate, which forms scales on the cells of one major group of oceanic phytoplankters, and the supporting structures of corals, and a further 2.5 gigatonnes is built into organic storage in wetlands and natural soils. This leaves about 4.4 gigatonnes to accumulate every

**Fig. 9.3** Experimental ponds being prepared at the University of Liverpool for experiments that showed that with increasing temperature, respiration increased much more than photosynthesis so that carbon stored in the sediments was lost as carbon dioxide to the atmosphere. A 4°C rise in temperature led to a 30% increase in carbon loss.

year in the atmosphere and explains most of the current temperature rise and associated disruption of previously experienced weather. Moreover we have converted about 75% of natural biomes to agriculture and pasturelands, which do not store much carbon, indeed often release it from previous soil stores through cultivation. The oceanic storage capacity is probably declining because the oceans are steadily acidifying through solution of newly released carbon dioxide, and acid waters prevent carbonate formation in algae and corals. Moreover, rising temperatures lead to increased respiration relative to photosynthesis in wetlands (Fig. 9.3) so that existing stores are prone to disappear. Cold waters dissolve a great deal of methane, produced from anaerobic respiration in sediments and peats, especially in the northern regions, but melting of the permafrost is beginning to release this and compound the problem. Methane is about twenty-three times more effective at retaining heat radiation than carbon dioxide.

## 9.4 Climate change in Britain and Ireland

There is no doubt that our climate is warming and that most of the change comes from increasing atmospheric greenhouse gas concentrations, resulting from our burning of fossil fuels. The expectations for western Europe are of continuing warming, greater total rainfall, more rain in winter and less in summer and a greater frequency of extreme events (torrential downpours, major floods, heat waves, droughts). There is ample evidence of an increasing trend in UK air temperatures, with the increase a little greater than the global average, though much lower than in the Boreal and Arctic. Long-term data series from world freshwaters show clear evidence for warming in streams and lake surface waters, and decreasing ice cover in northern and high-altitude lakes. Air temperature in central England has increased by about 1ºC since 1980 and there is a general correspondence between air and water temperatures where both are available.

Total rainfall in Britain has remained relatively steady since the 1760s, but the long-term trend in winter rainfall is of an increase (most prominent in Scotland and northern England) and a decrease in summer rainfall since 1874 in central England. There are large fluctuations, however, so that in the last fifteen years both trends have reversed, winters have been dry and summers very wet. There has been an increase in frequency of heavy rainfall events that are attributable to warming. Extreme winter flows and notable flooding have been more frequent over the past thirty years, particularly in the west and north. There is increasing confidence that these reflect climate changes. The frequency of droughts as yet shows no increase.

The organic content, reflected in the brown substances that are produced on decomposition of peat and wood, of our waters has been increasing across Europe since the 1980s, almost doubling in the last fifteen years. One view is that this has come from reduction in the acidity of rainfall (measures to combat acidification, see Chapter 8) or changes in land use (especially moorland management for grouse culture), but the trends in dissolved organic carbon are too consistent and widespread to be dependent on regional effects alone and ultimately climate change, directly or indirectly, is likely to be involved. Sea levels are also rising through expansion of the ocean volume and additional water coming through glacier melting. Coastal ponds and wetlands are likely to become saline as the sea overcomes them in the next few decades. This has already happened at Porlock in Somerset and the Norfolk and Suffolk Broads are particularly at risk.

## 9.5 Effects of warming

Temperature (subject to biogeographical accidents and barriers) has long been considered to determine the ranges of organisms. Climate change, hitherto, has confirmed this with northward and upward colonisation by many organisms in Britain. Overall the boundaries of 84% of over 300 species of aquatic bugs, dragonflies and damselflies, fish, amphibians, birds and mammals have moved northwards, two did not change and the rest have moved southwards. Stephen Thackeray and co-authors have analysed 25,532 rates of phenological change (timing of migrations, leaf appearance, flowering, mating, egg laying, hatching) between 1976 and 2005, for 726 UK terrestrial, freshwater and marine taxa. Many of the data have come from the records of amateur naturalists. Most spring and summer events have advanced rapidly and accelerated in a way that is consistent with observed warming trends, with the greatest changes after 1986. Advances in timing were overall lowest (2 days per decade) for predators, and about the same (4–4.5 days per decade) for primary producers (plants and algae) and grazers. The relative changes among plants, invertebrates and vertebrates differed among terrestrial, marine and freshwater habitats and there was no particular pattern among groups of finer taxonomic resolution. There may be some common patterns but these will only emerge with the collection of many more data.

There has long been an understanding that the sizes of warm-blooded animals (mammals and birds) tend to increase towards the Poles and this was attributed to the need to conserve heat in colder climates. Large animals have a smaller ratio of surface area to body mass than smaller ones. A similar relationship may exist for aquatic ectotherms (organisms whose temperature is determined by the environment rather than by internal metabolism). Bacteria, phytoplankton and a copepod placed in heated

**mesocosms**
experimental chambers
of large enough size
to contain a system
resembling a natural
one; see Fig.9.3

mesocosms (+2, +4 and +6 °C) significantly decreased in volume with temperature, and among wild fish stocks subjected to warming trends, mean size among species and mean size of individuals within species tends to decrease with temperature, whilst the proportions of small species and the proportions of juveniles tended to increase. A damselfly, *Erythromma viridulum*, has invaded England from the mainland continent since 1999. As it has moved northwards it has shown little genetic change but has increased in body size. Neither the reasons nor conservation significance of changes in size with warming are yet clear but it is claimed to be an emerging general principle. Size is important in food webs because the smaller an organism, the larger the range of grazers or predators that can take it. Here is an area ripe for investigations relating sizes of pond organisms to differences in mean temperature over a period of years.

Warming, indeed, is bringing in many new species from

more southerly latitudes. Some should be welcome as other cold-loving species disappear, but others, especially those introduced artificially, may become invasive and cause problems, at least temporarily until they become assimilated into the communities. About half of species introduced by human activity (for horticulture and to stock aquariums, for example) establish themselves in the wild and half of those spread. Warming may open up new opportunities for invasives, especially where high nutrient levels are found, as in most lowland freshwaters. This is because most invasion inevitably comes from warmer latitudes, where the biodiversity is greater, and because high nutrient levels favour competitive species that can build up high biomass. Fish, bird and plant introductions are generally soon noted but those of invertebrates may often be overlooked and those of non-clinical microorganisms are barely sought. There may be increasing potential for damaging invasions of invertebrates and although the reasons are several, climate change is almost certainly involved. Invasive freshwater shrimps (particularly *Dikerogammarus* species) are currently of concern because they are voraciously predatory.

New arrivals and establishment of breeding birds are generally welcomed. Almost uniformly they are birds with previous distributions to the south that visited Britain but did not breed, and thus warming is likely to be key to their establishment in Britain. Examples include little egret, common crane, whooper swan, Cetti's warbler, goldeneye, Mediterranean gull, great white egret, purple heron, cattle egret, pectoral sandpiper, spoonbill, red-necked grebe, little bittern, black-winged stilt, spotted sandpiper, little gull, bluethroat, black tern, ruff and Savi's warbler. Recent invaders that have begun to breed have first established themselves in protected areas such as Sites of Special Scientific Interest (SSSI). Such areas offer the highest quality habitats available and invasion of such places is unlikely by aggressive species likely to become pests. That is probably not true of exotic species among invertebrates and plants and probably not true of fish. They are likely to enter via estuaries from ship ballast water, and through aquaculture, water gardening, discarded live bait, and disenchanted aquarists, though also on the bodies or in the guts of migratory birds. Such species are often first found in canals, lowland rivers and ponds and reservoirs, all of them more challenging habitats than most SSSI. Such species are therefore likely to be strong competitors with a greater chance of becoming invasive problems.

It is likely that temperature effects *per se* will be less

important than effects of changed hydrology and that idiosyncratic behaviour of each species will lead to many indirect effects through biological interactions in communities. Freshwater organisms, however, are well adapted to disturbance and through invasion, redistribution, adaptation and microevolution will re-form functioning communities, though with probably very different composition from that at present. Some invasive species may come to dominate the new communities. These are interesting times for naturalists and their observations and records are extremely important in creating a picture of what is happening and in attempting to predict future changes, which is exceptionally difficult.

We are, however, proving reticent to curb carbon dioxide emissions from fuel burning and at present we need to about double our existing storages to stop temperatures rising. This will only happen when emissions are balanced by storage in sediments, peats and soils. The situation is on a scale previously unprecedented in human history. It is not one that tweaking individual national, even European, legislation can cope with. It has ramifications for everything we do, from the organisation of our societies to food production, population control, health, diets, and water usage. It may well be that we will not solve these problems and will have to take the consequences in chaos and disruption, but we would be well advised to try to solve them. And restoring former biomes with reflooding of wetlands, and re-creation of drained ponds and lakes would be to use nature's own solutions that have been continually tested. Some very large shallow lakes like the Aral Sea have been virtually lost, and former large lakes in the Fens and the Lancashire Plain have been completely drained. This would be safer rather than the exceptionally risky proposed geoengineering (launching reflective mirrors into space; increasing cloud formation by spraying sea water, for example) to counteract the symptoms, but not solve the problems. To be sure, even the current project of the Freshwater Habitats Trust (formerly Pond Action) to re-create a million ponds in the UK will have only a small absolute effect but it would be symptomatic of the change in attitude of mind that we urgently need. As the poet W.H. Auden wrote in 1947 (*Epilogue: The Age of Anxiety*):

*We would rather be ruined than changed. We would rather die in our dread than climb the cross of the moment and let our illusions die.*

# 10 Bibliography and further information

The Bibliography is arranged by sections. First I have given useful reference works for identification, then some general works and then more specific works appropriate to particular chapters. Some of these are papers in scientific journals that may need expensive subscriptions or access to a scientific library. However, papers are often available on other websites, for example University repositories or published under open access arrangements. A little exploration on the internet will often turn up a free copy. Where books are concerned, some are out of print. But again entry of the title into a search engine will usually turn up a set of second-hand booksellers offering used (pre-loved!) copies at low prices. No longer does one have to traipse round from shop to shop to find things. Finally I give some useful web sites and suppliers of equipment.

    G. Evelyn Hutchinson, in one of his books on limnology, headed the bibliography section with the biblical quotation 'Let us now praise famous men' (Ecclesiasticus 44.1). But you may join the ranks of people who publish. Writing up is an important part of a research project, which communicates the findings to other people. A really thorough, critical investigation that has established new information of general interest may be worth publishing if the organisms on which it is based can be identified with certainty. There are opportunities in the Newsletter of the Freshwater Biological Association and in a number of natural history journals, and, for material with an educational slant, the Journal of Biological Education. Contact with your local university, museum or natural history society, or the Natural History Museum in London will usually bring willing advice as to what is available. Then look at current numbers of appropriate journals to see what sorts of thing they publish, and then write a paper along similar lines. Keep it short, but present enough information to establish the conclusions. Consult an appropriate expert who can give advice on whether and in what form the material might be published. An unbreakable convention of scientific publication is that results are reported with scrupulous honesty. Hence it is essential to keep detailed and accurate records throughout the investigation, and to distinguish in the write-up between certainty and probability, and between deduction and speculation. It will usually be necessary to

apply appropriate statistical techniques to test the significance of the findings. A book such as Wheater & Cook (2003)
(see below) or *The OU Project Guide: Fieldwork and Statistics for
Ecological Projects* (Chalmers & Parker, 1989, available from the
Field Studies Council web site) will help, but this is an area
where expert advice can contribute much to the planning,
as well as the analysis, of the work. Hutchinson, the leading
ecologist of the twentieth century, wrote his first publication
(in 1918 in the *Entomologist's Record and Journal of Variation*
30, 138) based on observations of a swimming grasshopper
whilst he was aged 15 and still at school.

### Algae
(in order of comprehensiveness and general usefulness)
John, D.M, Whitton, B.A. & Brook, A.J. (2011) *The Freshwater Algal
Flora of the British Isles*. Natural History Museum and British
Phycological Society. 878 pp. *This is now the standard work for the
UK and will take you to species, except for diatoms. UK diatomists
think that the taxonomy is still in too much flux to produce a
standard key. There are, however, keys in German ,in several volumes
though expensive. If you can, find a used copy.*
Kelly, M. (2000) Identification of common benthic diatoms in
rivers. *Field Studies* 9 (4), 83–700. *A key to river diatoms, but many
also occur in ponds.* Available as an offprint from the Field
Studies Council website (www.field-studies-council.org/publi-
cations/aidgap-guides.aspx)
Belcher, J.H. & Swale, E. (1976) *A Beginner's Guide to Freshwater
Algae*. London: HMSO. Freely downloadable at www.bf.lu.lv/
grozs/HidroBiologjijas/Algae_quide.pdf
Belcher, J.H. & Swale E.M.F. (1979) *An Illustrated Guide to River
Phytoplankton*. London: HMSO. Freely downloadable at nora.
nerc.ac.uk/5243/1/Illustrated_river_phytoplankton.pdf. *Belcher
& Swale (1976, 1979) are particularly useful short guides with very
attractive illustrations. Second-hand originals can be bought, but the
downloaded scans are free.*
Canter-Lund, H. & Lund, J.W.G. (1995) *Freshwater Algae: Their
Microscopic World Explored*. Bristol: Biopress. *A lovely book of
photographs of algae and their protozoan and fungal parasites, and
interesting text.*
Moore, J.A. (1986) *Charophytes of Great Britain and Ireland*. BSBI
Handbook 5. London: Botanical Society of the British Isles. *A
key to charophytes, which has been updated and incorporated into
John, Whitton and Brook (2011).*
Hustedt, F. (1930) Die Süsswasser-Flora Mitteleuropas. Heft 10:
Bacillariophyta (Diatomeae). Jena, Germany: Gustav Fischer.
*This has been replaced by very expensive multi-volume German
diatom floras that give much more information on variation and
updated nomenclature, and which are themselves being put out of
date by revisions of diatom taxonomy, but for genera and many
prominent species this small book is hard to beat. It is scarce and
in German, but compact and very well illustrated. Revisions in
taxonomy can always be traced through AlgaeBase (British Phycological Society) on the internet.*

## Protozoa

Both of the first two are useful but new copies are expensive.Hingley (1993) specialises on organisms in *Sphagnum*, including protozoans.

Jahn, T.L., Bovee, E.C. & Jahn, F.F. (1978). *How to Know the Protozoa*. New York: McGraw Hill.

Patterson, D.J. (2003) *Free-living Freshwater Protozoa: a Color Guide*. Washington: ASM Press.

Hingley, M. (1993) *Microscopic Life in* Sphagnum. Naturalists' Handbooks 20. Slough: The Richmond Publishing Co. Ltd.

## Aquatic plants

(in order of comprehensiveness and general usefulness)

Stace, C. (2010) *New Flora of the British Isles*. Third edition. Cambridge: Cambridge University Press. *The standard flora now for plants of the British Isles. It is published as a main work and as an excursion flora. It is a professional key demanding use of botanical terminology (there is a glossary) and without pictures.*

Haslam, S., Sinker, C.A. & Wolseley (1982) *British Water Plants*. Field Studies Council. *An easily usable key to most of the flora, available from the Field Studies Council website.*

Rose, F. (updated by C. O'Reilly) (2006) *The Wildflower Key (Revised Edition)*. London: Warne. *Less comprehensive than Stace, but with pictures. There are several other illustrated wild flower keys of the same type.*

Fitter, R., Fitter, A. & Farrer, A. (1984) *Collins Pocket Guides. Grasses, Sedges, Rushes and Ferns of Britain and Northern Europe*. London: Collins . *Many emergents are in this group and are not well covered in wildflower books.*

Preston, C.D. & Croft, J.M. (1997) *Aquatic Plants in Britain and Ireland*. Colchester: Harley Books. *Not a key, but a set of descriptions and distribution maps.*

*Sphagnum*. A free key is downloadable at www.bbsfieldguide. org.uk/sites/default/files/pdfs/otherpdfs/BBS%20Field%20 Guide%20Sphagnum%20Key.pdf . It includes most of the British species.

## Invertebrates

(in order of comprehensiveness and general usefulness)

Dobson, M., Pawley, S., Fletcher, M. & Powell, A. (2012) *Guide to Freshwater Invertebrates*. Freshwater Biological Association Special Publication 68. Ambleside: Freshwater Biological Association. *This is the first port of call. It will direct you to more detailed keys in many instances. A lot are still available from the Freshwater Biological Association online shop at www.fba.org.uk/ shop/. Those out of print may be available for sale on the web.*

*Another useful source is the Field Studies Council* (www.field-studies-council.org/publications/aidgap-guides.aspx) which produces the AIDGAP series of keys to stoneflies, water plants, freshwater invertebrates, freshwater fish, benthic diatoms, freshwater bivalves, caddis larvae, caddisflies and mayflies and also Royal Entomological Society keys to water beetles, the Synopses of British Fauna series (freshwater ostracods) and easily usable fold-out elementary charts to dragonflies and damselflies, wetland birds, fish and *Sphagnum*.

Donner, J. (1966) *Rotifers*. Translated by G.S. Wright. London: Frederick Warne. *A useful book on a group for which the literature is scattered.*

Guthrie, M. (1989) *Animals of the Surface Film*. Naturalists'
Handbooks 12. Slough: The Richmond Publishing Co. Ltd. *A
specialised but accessible guide to one aspect of ponds.*

## Fish
Maitland, P.S. (2004) *Keys to the Freshwater Fish of Britain and
Ireland, with Notes on their Distribution and Ecology*. Freshwater
Biological Association Scientific Publication 62. Ambleside:
Freshwater Biological Association. *There are several easily usable
guides to fish but this is compact and authoritative.*

## Other vertebrates
Beebee, T. (2013) *Amphibians and Reptiles*. Naturalists' Handbooks
31. Exeter: Pelagic Publishing.
*There are many bird guides that are readily available and which
variously appeal to different tastes.*

## General works on ecology
Begon, M., Townsend, C. & Harper, J.L. (2005) *Ecology: From
Individuals to Ecosystems*. Fourth Edition. Chichester: Wiley. *An
excellent general text on ecological principles.*
Burgis, M.J. & Morris, P. (2007) *The World of Lakes: Lakes of
the World*. Special Publications of the Freshwater Biological
Association 15. Ambleside: Freshwater Biological Association.
*Written for a lay audience and covering the world's lakes with
attractive illustrations.*
Clegg, J. (1965) *The Freshwater Life of the British Isles*. Wayside and
Woodland Series. London: Frederick Warne. *A general work, full
of useful information on particular animals and plants.*
European Pond Conservation Network (2008) *Manifesto*. Down-
loadable at www.europeanponds.org *The EPCN network works
to conserve pond systems and this Manifesto describes the problems.*
Finlay B.J. & Maberly S.C. (2000) *Microbial Diversity in Priest Pot. A
Productive Pond in the English Lake District*. Ambleside: Freshwater
Biological Association. *Priest Pot is a very small lake, which has had
a great deal of attention. The protozoans and algae have been extensively
studied. Lists of all the organisms yet identified in it are given.*
Fitter, R. & Manuel, R. (1994) *Lakes, Rivers, Streams and Ponds*.
Collins Photo Guide. London: Harper Collins. *A comprehen-
sive general guide with good photographs or line diagrams of the
commoner species.*
Fryer, G. (1991) *A Natural History of the Lakes, Tarns and Streams
of the English Lake District*. Ambleside: Freshwater Biological
Association. *Lots of attractive drawings and a hand-lettered text by
one of the most knowledgeable freshwater biologists and naturalists of
recent decades.*
Kabisch, K. & Hemmerling, J. (1984) *Ponds and Pools: Oases
in the Landscape*. Translated from the original German by
Ilse Lindsay. London: Croom Helm. *Packed with ecological
information and covers well the general principles.*
Macan T.T (1973) *Ponds and Lakes*. London: Longmans. *A well
written classic that concentrates particularly on the ecology of the
invertebrates.*
Macan, T.T. (1963) *Freshwater Ecology*. London: Longmans, Green
and Co. *A detailed, fairly heavy treatment, concentrating on invertebrates and their ecology.*
Mellanby, H. (1963) *Animal Life in Freshwater*. London: Methuen.

*A small, straightforward book that systematically recounts the biology of all the freshwater invertebrate groups. A must.*

Moss, B. (2010) *Ecology of Freshwaters, Fourth Edition, A View for the Twenty-First Century.* Chichester: Wiley. *The current most comprehensive book on freshwater ecology in general.*

Moss, B. (2012) *Liberation Ecology: the Reconciliation of Natural and Human Cultures.* Excellence in Ecology 24. Oldendorf/ Luhe, Germany: International Ecology Institute. *An account of the major principles of ecology and environment, intended for a non-scientist audience and approached through the media of fine art, music and literature.*

Moss, B. (2015) *Lakes, Loughs and Lochs.* London: Collins New Naturalist. *A New Naturalist book concentrating on Britain and Ireland and written for a general audience.*

Purseglove, J. (1988) *Taming the Flood.* Oxford: Oxford: University Press. *Deals with river engineering and floodplain drainage and their consequences. Passionate and beautifully illustrated.*

Thompson, G., Coldrey, J. & Bernard, G. (1985). *The Pond.* London: Collins. *Short on text but a collection of excellent photographs of pond organisms mostly in situ.*

Wheater, C.P. & Cook, P. (2003) *Studying Invertebrates.* Naturalists' Handbooks 28. Slough: The Richmond Publishing Co. Ltd. *Contains much information on sampling and use of simple statistics.*

Williams, P.J., Biggs, J., Whitfield, M., Thorne, A., Bryant, S., Fox, G. & Nicolet, P. (2010) *The Pond Book: A Guide to the Management and Creation of Ponds. 2nd Edition,* eds. Williams, P. and Julian, A.M. Oxford: Pond Conservation. Now available from Freshwater Habitats Trust, Oxford (freshwaterhabitats.org.uk/ habitats/pond/pond-book/). *Another must. A wealth of practical information.*

## Chapter 1

Bennion, H., Harriman, R. & Battarbee, R. (1997) A chemical survey of standing waters in south-east England with reference to acidification and eutrophication. *Freshwater Forum,* 8, 28–44

Cereghino, R., Biggs, J., Oertli, B. & Declerk, S. (2008) The ecology of European ponds: defining the characteristics of a neglected freshwater habitat. *Hydrobiologia* 597, 1–6.

De Meester, L., Declerck, S., Stoks, R., Louette, G., Van de Meutter, F., De Bie, T., Michels, E. & Brendonck, L. (2005) Ponds and pools as model systems in conservation biology, ecology and evolutionary biology. *Aquatic Conservation: Marine and Freshwater systems* 15, 715–725.

Jeffries, M. (2011) Ponds and the importance of their history: an audit of pond numbers, turnover and the relationship between the origins of ponds and their contemporary plant communities in south-east Northumberland, UK. *Hydrobiologia* 689, 11–21.

Oertli, B., Cereghino, R., Hull, A. & Miracle, R. (2009) Pond conservation: From science to practice. *Hydrobiologia* 634, 1–9.

Sutcliffe, D.W. (1998) The ionic composition of surface waters in the English Lake District, related to bedrock geology, with some singular facts and speculation on the existence of mineral-rich groundwaters. *Freshwater Forum* 11, 30–51.

Verpoorter, C., Kutser, T., Seekell, D.A. & Tranvik, L.J. (2014) A global inventory of lakes based on high-resolution satellite imagery. *Geophysics Research Letters* 41, 6396–6402.

Williams, P., Biggs, J., Fox, G., Nicolet, P. & Whitfield, M.

(2001) History, origins and importance of temporary ponds. *Freshwater Forum* 17, 7–15.

## Chapter 2

Bronmark, C. & Hansson, L.-A. (Eds) (2012) Chemical Ecology in Aquatic Systems. Oxford: Oxford University Press. *High level and deals with chemical communication among animals in the water, but mainly marine.*

Pennak, R.W. (1985) The fresh-water invertebrate fauna: Problems and solutions for evolutionary success. *American Zoologist* 25, 671–87. *A very readable account.*

## Chapters 3 and 4

Cook, J, Chubb, J.C. & Veltkamp, C.J. (1998) Epibionts of *Asellus aquaticus* (L.) (Crustacea, Isopoda): and SEM study. *Freshwater Biology* 39, 423–438. *Illustrates just how complex and intricate, relationships can be in pond animals.*

Finlay, B.J. & Esteban, G.F. (1998) Freshwater protozoa: biodiversity and ecological function. *Biodiversity and Conservation* 7, 1163–1186. *A very useful general account, but uses the older classification system before new information became available on the organisation of kingdoms.*

Walker, G., Dorrell, R.G., Schlacht, A. & Dacks, J.B. (2011) Eukaryotic systematics: a user's guide for cell biologists and parasitologists. *Parasitology* 138, 1638–63. *An account of the new classification of kingdoms. Hard going.*

Whittaker, R.H. (1969) New concepts of the kingdoms of organisms. *Science* 163, 150–60. *A classic of its time and still interesting to read.*

## Chapter 5

Cribb, S., Cribb, J. (1998) *Whisky on the Rocks.* Edinburgh: British Geological Survey.

Elser, J.J., Bracken,M. E.S.,, Cleland, E.E., Gruner, D.S., Harpole, W.S., Hillebrand, H., Ngai, J.T., Seabloom, E.W., Shurin, J.B. & Smith, J.E. (2007) Global analysis of nitrogen and phosphorus limitation of primary producers in freshwater, marine and terrestrial systems. *Ecology Letters* 10, 1135–1142.

## Chapter 6

Forbes, S. (1887) The lake as a microcosm. *Bulletin of the Scientific Association (Peoria, IL)* 1887, 77–87. *An important early paper that set the scene for the development of freshwater science.*

Gill, J.L., Williams, J.W., Jackson, S.T., Lininger, K.B. & Robinson, G.S. (2009) Pleistocene megafaunal collapse, novel plant communities, and enhanced fire regimes in North America. *Science* 326, 1100–1103.

Hutchinson, G.E. (1959) Homage to Santa Rosalia or why are there so many kinds of animals? *American Naturalist* 93, 145–59. *A classic in the development of niche theory.*

Lindeman, R.L. (1942) The trophic-dynamic aspect of ecology. *Ecology* 23, 399–417. *Another classic that developed the main ideas of food webs and the importance of detritus in lakes. The previous four papers are widely available for free on the Web.*

MacArthur, R.H. & Wilson, E.W. (1967, reprinted 2001) *The Theory of Island Biogeography.* Princeton: Princeton University Press.

Moss, B. (2015) Mammals, freshwater reference states, and the mitigation of climate change. *Freshwater Biology* 60, 1964–1976.

*An example of how new ideas can emerge from casual observations.*

Odum, E.P. (1969) The strategy of ecosystem development. *Science* 164, 262–270.

Oertli, B., Joye, D.A., Castella, E., Juge, R., Cambin, D. & Lachavanne, J.-B. (2002) Does size matter: The relationship between pond area and biodiversity. *Biological Conservation* 104, 59–70.

## Chapter 7

Boll, T., Johansson, L.S., Lauridesen, T.L., Landkildehus, F., Davidson, T.A., Søndergaard, M., Andersen, F.O. & Jeppesen, E. (2012) Changes in benthic macroinvertebrate abundance and lake isotope (C, N) signals following biomanipulation: an 18-year study in shallow Lake Vaeng, Denmark. *Hydrobiologia* 686, 135–145.

Brooks, J.L. & Dodson, S.I. (1965) Predation, body size and composition of plankton. *Science* 150, 28–35.

Giles, N. (1992) *Wildlife after Gravel: Twenty Years of Practical Research by the Game Conservancy and ARC.* Fordingbridge: Game Conservancy. *Ecological studies in restored gravel pit ponds.*

Hurlbert, S.H., Zedler, J. & Fairbanks, D. (1972) Ecosystem alteration by mosquitofish (*Gambusia affinis*) predation. *Science* 175, 639–641.

Ings, N.I., Hildrew, A.G. & Grey, J. (2012) House and garden: larval galleries enhance resource availability for a sedentary cadddisfly. *Freshwater Biology* 57, 2526–2538.

Scheffer, M., Carpenter, S., Foley, J.A. & Walker, B. (2001) Catastrophic shifts in ecosystems. *Nature* 413, 59–596.

Scheffer, M. (2009) *Critical Transitions in Nature and Society.* New Jersey: Princeton University Press. *The wider application of alternative states ideas.*

## Chapter 8

Andrews, J. & Kinsman, D. (1990). *Gravel Pit Restoration for Wildlife.* Sandy: Royal Society for the Protection of Birds.

Aston, S. (1988) *Mediaeval Fish, Fisheries and Fishponds in England. British Archaeological Research British Series* 182. Oxford: British Archaeological Society.

Bulleid, A. (1924, revised 1958). *The Lake-Villages of Somerset.* Glastonbury: Glastonbury Antiquarian Society.

Johnes, P., Moss, B. & Phillips, G. (1996) The determination of total nitrogen and total phosphorus concentrations in freshwaters from land use, stock headage and population data: testing of a model for use in conservation and water quality management. *Freshwater Biology* 36, 451–473. *The use of desk studies to calculate approximate nutrient budgets for catchments.*

Moore, N.V. (2002) *Oaks, Dragonflies and People.* Leiden: Brill Books. *An account of the construction, after his retirement, of a pond specifically to support a high diversity of dragonflies by one of the most prominent of twentieth century conservation scientists.*

Williams, P., Whitfield, M. and Biggs, J. (2008) How can we make new ponds biodiverse? A case study monitored over 7 years. *Hydrobiologia* 597, 137–148.

Wood, P.J., Greenwood, M.T. & Agnew, M.D. (2003) Pond diversity and habitat loss in the U.K. *Area* 35, 206–216

**Chapter 9**

Biggs, J., Williams, P., Whitfield, M., Nicolet, P. & Weatherby, A. (2005) 15 years of pond assessment in Britain: results and lessons learned from the work of Pond Conservation. *Aquatic Conservation: Marine and Freshwater Ecosystems* 15, 693–714.

Cole, J., Prairie, Y.T., Caraco, N.F., McDowell, W.H,, Tranvik L.J., Striegl, R.G., Duarte, C.M., Kortelainen, P., Downing, J.A., Middelburg, J.J. & Melack, J. (2007) Plumbing the global carbon cycle: Integrating inland waters into the terrestrial carbon budget. *Ecosystems* 10, 171–84. *The role of small lakes, ponds and wetlands in the carbon cycle. Heavy going.*

Food and Agriculture Organization of the United Nations (2014). *The State of World Fisheries and Aquaculture.* Rome: FAO. Downloadable at www.fao.org/3/a-i3720e.pdf. *The current situation of fisheries.*

Hickling, C.F. (1971) *Fish Culture.* London: Faber and Faber. *An account packed with interesting detail and concentrating particularly on fish culture in the tropics.*

Lenton, T. & Watson, A. (2011). *Revolutions That Made the Earth.* Oxford: Oxford University Press. *An account of Earth's somewhat dramatic ecological history.*

Lovelock, J. (2000) *Gaia: A New Look at Life on Earth.* 2nd Edition. Oxford: Oxford Paperbacks. *A classic that examines how atmospheric and ocean composition are maintained within equable limits for life. The book has caused a great deal of controversy and vehement rejection by some evolutionary biologists on the basis of one romantic but probably wrong idea in it that caught the imagination of the environmental movement. However, it was well received by earth chemists, oceanographers and geologists. You should read it because of the controversy and extreme reaction, but also read Lenton & Watson (2011) and Tyrell (2013), which gives a balanced examination of the evidence, rejects the Gaia hypothesis of an Earth superorganism but also clearly establishes that Lovelock raised important issues and made a major contribution.*

Moss, B. (2010) Climate change, nutrient pollution and the bargain of Dr Faustus. *Freshwater Biology* 55, 171–183. *One of two accounts (see also Yvon-Durocher et al. (2010) published in the same year showing that rising temperatures are likely to increase dramatically the carbon dioxide content of the atmosphere through increased respiration of stored carbon.*

Moss, B. (2014) Freshwaters, climate change and UK conservation. *Freshwater Reviews* 7, 25–75. *A comprehensive review of what is happening.*

Raebel, E.M. Merckx, T., Feber, R.E., Riordan, P., MacDonald, D.W. & Thompson, D.J. (2012) Identifying high-quality pond habitats for Odonata in lowland England: implications for agri-environment schemes. *Insect Conservation and Diversity* 5, 422–432.

Sayer, C., Andrews, K., Shilland, E., Edmonds, N., Edmonds-Brown, R., Patmore, I., Emson, D. & Axmacher, J. (2012) The role of pond management for biodiversity conservation in an agricultural landscape. *Aquatic Conservation: Marine and Freshwater Ecosystems* 22, 626–638.

Thackeray, S.T. and 24 others (2010) Trophic level asynchrony in rates of phenological change for marine, freshwater and terrestrial environments. *Global Change Biology* 12, 3304–3313.

Tyrrell, T. (2013) *On Gaia: A Critical Investigation of the Relationship between Life and Earth.* Princeton, USA: Princeton University Press.

Watts, G. and 24 others (2015) Climate change and water in the UK-past changes and future prospects. *Progress in Physical Geography* 39, 6–28.

Yvon-Durocher, G., Jones, J.I., Trimmer, M., Woodward, G. & Montoya, J.M. (2010) Warming alters the metabolic balance of ecosystems. *Philosophical Transactions of the Royal Society* B 365, 2117–2126. *See also Moss (2010).*

## Useful websites

Freshwater Habitats Trust (Pond Action). This organisation is an enthusiastic proponent for pond conservation and creation. It has a number of schemes in which amateurs can get involved and an archive of useful articles for downloading. Its website is well worth an exploration: freshwaterhabitats.org.uk

British Geological Survey (Maps, Regional Handbooks) www.bgs.ac.uk/catalogue/home.html

Environment Agency River Basin Management Plans www.gov.uk/government/collections/river-basin-management-plans#current-river-basin-management-plans

Natural Resources Wales River Basin Management Plans naturalresources.wales/water/quality/submission-of-river-basin-management-plans/?lang=en

Scottish Environment Protection Agency River Basin Management Plan www.sepa.org.uk/media/28300/scotland_rbmp_sea_environmental_report.pdf

## Equipment suppliers

Brunel microscopes (www.brunelmicroscopes.co.uk). General supplies for microscopy including Naphrax diatom mountant.

Duncan Associates (www.duncanandassociates.co.uk). This company makes higher quality equipment for freshwater sampling than that offered by general scientific suppliers. It is especially worth buying high quality pond and plankton nets.

Fisher Scientific (www.fisher.co.uk). A general supplier of equipment.

Hanna Instruments (www.hannainstruments.co.uk). Hanna make excellent pocket conductivity probes at a very reasonable price. See: www.hannainstruments.co.uk/conductivity-tds.html

# Index

Page numbers in *italics* denote figures and in **bold** denote tables.